KU-017-530

Data analysis for biomolecular sciences

Jon Maber
University of Leeds

LONGMAN

Pearson Education Limited
Edinburgh Gate, Harlow
Essex CM20 2JE
England

and Associated Companies throughout the World

© Pearson Education Limited 1999

The right of Jon Maber to be identified
as author of this Work has been asserted by him in
accordance with the Copyright, Design and
Patents Act 1988.

All rights reserved; no part of this publication may be
reproduced, stored in any retrieval system, or transmitted
in any form or by any means, electronic, mechanical,
photocopying, recording, or otherwise without either the prior
written permission of the Publishers or a licence permitting
restricted copying in the United Kingdom issued by the
Copyright Licensing Agency Ltd., 90 Tottenham Court Road,
London W1P 9HE.

Screen shots reprinted by permission from Microsoft Corporation.

First published 1999

ISBN 0 582 305950

British Library Cataloguing-in-Publication Data
A catalogue record for this book is
available from the British Library.

Library of Congress Cataloging-in-Publication Data
A catalog entry for this title is available
from the Library of Congress.

Set by 35 in 9/12 pt Stone Serif
Printed in Great Britain by Henry Ling Ltd., at the Dorset Press,
Dorchester, Dorset.

Contents

Preface

···

This book deals with the type of statistics that is used to analyse data from biological tests and assays. It also goes beyond the mathematical aspects of data handling to put the subject in the context of policy making and problem solving. The text deals with fundamental ideas that are applicable to many employment situations but most of the case studies and examples are drawn from the work of scientists who assay biological substances. This approach makes the book particularly relevant to students of biochemistry, physiology, pharmacology and other related subjects.

Scientists need skills and knowledge in the areas of mathematics, English and IT, and undergraduates entering science courses are expected to have good qualifications in each of these areas. However, few students at the start of their studies have much experience of combining these three strands in a scientific context. Students at school are rarely asked to explain what a calculation means in a mathematics examination, and in English examinations the subject of essays and comprehension questions is rarely technical. Even in science examinations, questions tend to be segregated into those that involve explanation and those that involve calculation. This text attempts to put every statistical and mathematical idea into its proper scientific context and emphasizes the use of mathematics and statistics for communicating ideas and making decisions.

How to use this book

Most chapters assume that you have studied previous chapters and understood the ideas there. This progression is unavoidable because so many statistical ideas have been built on other statistical ideas. Here is a summary of what previous knowledge is needed for effective study of each chapter.

Chapter 1 – Probability

This chapter introduces probability and assumes only a basic knowledge of arithmetic and a grasp of logical statements in English. Probability concepts underpin the ideas raised in all the other chapters and it is worth studying this chapter for revision even if you have previous knowledge of the subject. You may like to jump directly to Chapter 7 which presents a case study if you are interested in how you would use probability methods in a policy-making situation.

Chapter 2 – Significance

This chapter is about the way statistics can be used to decide whether a hypothesis is true or false based on systematic analysis of data. Since probability ideas run through this chapter, a good understanding of the ideas presented in Chapter 1 is important. The significance test presented in this chapter is explained in detail, particularly where general principles are applicable to the other types of significance test covered in later chapters.

Chapter 3 – Measurements and assays

This chapter introduces the idea that random and systematic errors affect measurements and assays. The random element in measurements means that probability is involved. Some explanation of mathematical operations involved in using assays is presented and a basic ability in algebra is assumed.

Chapter 4 – Precision

This chapter looks in detail at the precision of assays. It assumes that you know what an assay is and why imprecision can occur, and because it goes on to cover the mathematics involved in quantifying random variation, a thorough understanding of probability is required.

Chapter 5 – Accuracy

An understanding of all the previous chapters is important before you study this chapter. Quantifying the accuracy of an assay is made more complex because it is also imprecise, and determining whether an assay is inaccurate involves the use of a significance test. The general principles of significance testing are not repeated in this chapter, only the specifics of the t test and AnoVa. A basic understanding of algebra is assumed and also the algebra of logarithms presented in Chapter 3.

Chapter 6 – Relationships between variables

When you study the relationships between two variables you have to make measurements and you have to take into account the imprecision of those measurements. The doubt about the data due to imprecision may mean that you also have to quantify the significance of any relationship. Consequently, an understanding of ideas in all the previous chapters is needed.

Chapter 7 – Screening for colorectal cancer

This chapter assumes an ability in basic arithmetic and logical English and an understanding of probability as presented in Chapter 1. You can choose to study this chapter ahead of the others if you would like to see how data handling fits into decision and policy making.

Chapter 8 – Data handling with a computer

This chapter reiterates some of the ideas in previous chapters but in the context of doing calculations with a computer. Worked examples using Microsoft Excel are included. You might like to dip into sections of this chapter as you study the previous chapters or you could use this chapter as reference material as you do computer-based exercises from Chapter 9.

Chapter 9 – Exercises

The exercises in this chapter are grouped in sections corresponding to the other chapters of the book. It would probably be best to attempt the appropriate exercises after you have studied each chapter.

Spelling variations

This book is written with international English spelling. American readers should take note of the following spelling variations:

International	American
Faeces	Feces
Faecal	Fecal
Foetus	Fetus
Foetal	Fetal
Haem	Heme
Programme	Program (The American spelling is used internationally when referring to computer programs but not in other contexts.)

Greek letters

A number of letters from the Greek alphabet are traditionally used in statistics and in science. The Greek letters used in this text are:

Letter	Pronunciation	Use in this text
α	alpha	In the name of a protein
χ	ky, to rhyme with sky	A traditional symbol for a statistical value
μ	mu	Traditional symbol for population mean and also stands for the number 10^{-6}, when it is pronounced 'micro'
σ	sigma	Traditional symbol for population standard deviation
Σ	capital sigma	Mathematical symbol that indicates the sum of a number of terms

Chapter 1

Probability

Introduction

Probability is the branch of mathematics that quantifies uncertainty. This quantification is done for a purpose: to reliably communicate levels of uncertainty or to enable decision making. This chapter will advocate the use of effective language combined with proper analysis of data, show the importance of probability calculations for problem solving and decision making, explain basic probability calculations and look at common misconceptions in the area of probability. A detailed worked example is presented in Chapter 7.

Making medical decisions involves uncertainty

Medics are required to make decisions concerned with the diagnosis and treatment of patients throughout their careers, and scientists in the field of medicine can contribute to decisions that affect health policy for entire populations. These decisions may be based on information about patients, diseases, medical tests and treatments and there is uncertainty to some degree in all these things. Let's take some specific examples.

Variation in the symptoms for a disease

Not every individual with a particular disease suffers the same symptoms to the same degree. For example, diabetes may be accompanied by retinopathy (damage to the retina) or it may not. This sort of variation may make the process of diagnosis more difficult.

Blood glucose test

There is some small uncertainty about the concentration of blood glucose determined by a laboratory; the number provided to the doctor may be a little higher

1

or a little lower than the correct value. The assay for blood glucose is reliable enough to allow decisions to be made by doctors. For example, the normal venous blood serum glucose concentration for a fasting adult lies in the range 7–10 mg ml^{-1}, and if the level of blood glucose in your patient is 2 mg ml^{-1}, an error of 0.1 mg ml^{-1} doesn't much matter. This person has something wrong with them and more investigation is needed.

Alpha-foetoprotein assay

Alpha-foetoprotein is a protein found in minute quantities in adult blood. Its concentration can rise in certain medical situations and so measuring it can be important. Because the protein has no enzymatic activity or distinctive chemical composition and because its concentration may be as low as one thousand millionths of a gram per millilitre, it is difficult to measure with sufficient precision and accuracy. So there can be much more doubt about the results. Until recent decades the assay for this protein was not reliable enough to base any medical decisions on. We will return to the α-foetoprotein assay in later chapters.

How can probability help?

There are three advantages to a quantitative approach to uncertainty.

1. It allows the reliable communication of uncertainty

It is difficult to communicate in ordinary language the probability of an event occurring. You're forced to use phrases like 'highly likely', 'not much chance', 'almost certain'. These could be interpreted in different ways by different people. Imagine that you are a doctor who wants to treat a patient and you are asking a consultant about a new treatment. If she says it's 'fairly likely to work', will you understand that the way she intended? It might depend on the inherent optimism or pessimism of both you and her. It would help to quantify the probability. If the consultant says 'there is a 95% chance that the treatment will work', there is no room for misinterpretation.

This aspect of probability doesn't rule out the use of ordinary human judgement by doctors but it is an attempt to eliminate the influence of ambiguous English and personal philosophy on the doctor's decision making. When the perception of a level of uncertainty is passed from person to person via the medium of ordinary language it can be modified and becomes less and less useful. But when a number is used to measure an uncertainty it is easier to communicate that number and keep its meaning unchanged and unambiguous. Ideally, when a probability is communicated there will also be an explanation of what data were used to calculate it and how those data were collected. This ensures that a decision maker can see if the probability was determined from a study that is applicable to their own situation.

2. It allows more informed decision making

Clinical decisions are often made based on the results of tests. There can be uncertainty about many test results because of natural biological variation between individual patients or because the test method itself is unreliable. If a medical test must be unreliable, it's still possible to quantify that unreliability so that this can be weighed against the consequences of different decisions. A naive approach to unreliable information is to discard it and demand reliable information, but in many situations the choice is between unreliable information and no information at all.

For example, a test for blood in the faeces can be made as part of a routine health check. The aim of the test is to detect cancer of the colon. However, this is a very uncertain test since the probability that the individual receiving a positive result actually has cancer of the colon may be only 3%. (There are many other causes of blood in the faeces.) Would you throw away results from this test and ignore them? How can the test help a doctor make a decision? Would you send this patient for an operation on the colon based on a test that is 3% reliable? With this particular test a positive result will not lead to a decision to operate but will lead to a decision to view the inside of the patient's colon with an endoscope. A more detailed examination of this test is presented in the case study in Chapter 7.

3. It enables a systematic approach to decision making

There is frequently a degree of uncertainty about medical treatments, so decisions about which treatment to use with a particular patient often involve subjective judgement. The people who make such decisions will find that they repeat the same decision-making processes throughout their careers in case after case although each case is unique in terms of the information that is available. For example, a consultant in the field of oncology may have to decide whether to operate on a patient with a breast tumour. This will be an individual decision based on the information available about this patient's case, but the decision-making process is likely to be similar for all the consultant's other patients.

There is room for a systematic approach to decision making in some medical situations. The science of probability can't take away the need for personal judgement among clinical staff but it can help them to apply judgement. Judgement that is based on systematically quantified risk and benefit assessments, using probability calculations, is more reliable and ultimately will benefit the health of patients. This approach doesn't make life easy for medics – in many ways it is more challenging because one must be familiar with the most recent and reliable statistics and be able to constantly review the validity of the probability figures against the information available for ones own patients. One must also make very difficult comparisons of pros and cons, for example, balancing the probability that a treatment will work against the consequences of its failure.

How difficult is probability?

Probabilities cause some people difficulties. This is perplexing since the mathematics is often very simple. Perhaps one reason is that the human animal is equipped from birth with an ability to survive by making judgements about events, based on experience. Human judgement often fails in less 'natural' situations. So getting to grips with probability can be more a process of 'unlearning' your misconceptions than learning new things. Another difficulty comes when different probabilities are combined because they require the strict logical use of the English language, in particular the exact use of the words 'not', 'and' and 'or' in statements of fact.

Probability is a combination of mathematics and English

This may be something that surprises you, but a probability is a number which is always connected to a statement in English (or some other language which can make statements of fact). A statement in English that is absolutely true needs no number with it, and a statement which is absolutely false only needs a slight rephrasing to be made into a statement that is absolutely true.

A true statement needs no numbers:

The capital of England is London.

A false statement such as

The capital of England is Paris.

can be expressed as a true statement using the word 'not' (again no numbers needed):

The capital of England is NOT Paris.

The science of probability still uses a statement in English but adds a number to indicate the level of uncertainty on a scale from false to true.

An uncertain statement:

A patient who receives a positive result from a faecal occult blood test has cancer of the colon.

can be expressed using a probability:

A patient who receives a positive result from a faecal occult blood test has a 3% chance of having cancer of the colon.

How would you cope with this if you couldn't use numbers? Here's a different version of the previous statement:

A patient who receives a positive result from a faecal occult blood test may have cancer of the colon but it's fairly unlikely.

Or if you wanted to influence doctors away from using the test:

A patient who receives a positive result from a faecal occult blood test won't have cancer of the colon unless they are unlucky.

Someone with a religious world view might even phrase it in the following way:

A patient who receives a positive result from a faecal occult blood test will only have cancer of the colon if such is the will of god.

These alternatives may all be valid expressions of the writer's opinion but they each depend on the reader having a similar understanding of the wording used and a similar philosophical outlook. The statement that uses a number involves English that is not open to interpretation. The element of uncertainty is expressed using a number which again is not open to interpretation by the reader. Of course, different people might still take different action based on this information but at least they can all work from the same reference point before adding their personal judgement.

Probability conventions

There are several alternatives for representing probabilities. They are all fractions of one sort or another that indicate the position of a statement on a scale from false to true. Which convention you choose will depend on who you want to communicate the probability to. Here is a summary of the alternatives.

Mathematical

Mathematicians prefer to use a number between 0.0 and 1.0 to represent probability. An event with a probability of 0.0 can never happen; an event with a probability of 1.0 must happen; and an event with a probability of 0.5 is just as likely to happen as not happen. (More generally, 0.0 indicates a statement that is false, 1.0 indicates a statement that is true and 0.5 indicates a statement that is equally likely to be true as to be false.) This form of probability is the easiest to manipulate when you make calculations but will be unfamiliar to non-mathematicians.

The probability that the next throw of a coin will come up heads is 0.5. If you isolate the statement, it is 'The next throw of a coin will come up heads', to which we have attached a probability of 0.5.

Percentage

A common way to quote probabilities in reports for non-mathematicians is to use a percentage. An event with a probability of 0% can never happen; an event with a probability of 100% must happen; and an event with a probability of 50% is just as likely to happen as not happen. Percentages remind us that the figure is a proportion of certainty.

Integer fractions

Sometimes you need to communicate ideas of uncertainty to people who are not only non-mathematicians but are barely numerate and find fractional numbers difficult to understand. This is the reason that gambling odds given by betting offices are not given as percentages but as the ratio of two whole numbers. So, if an event is as likely to occur as not occur then the probability of it occurring is one in two. Let's say that we will toss a coin. The probability of heads is 0.5 or 50%. Despite the fact that we will toss the coin only once we can say the probability of heads is one in two. It's stated as if we were intending to toss the coin twice. Although the 0.0 to 1.0 and the 0% to 100% scales are incomprehensible to many people, the 3 in 1000 form of probability can still cause great confusion.

For example, if you say:

'The probability that rolling two dice will score twelve is 1 in 36.'

you may get the reply:

'You've rolled the dice 35 times and not scored twelve yet so this time it must score twelve.'

When we say 1 in 36 we are trying to express a fraction and not indicating a specific number of trials. If we wanted to avoid this specific comment we might quote the probability as 100 in 3600. We can't say exactly how many heads there would be for a specific number of trials, so using big numbers just makes it difficult to see what the proportion is. If you have to communicate probability to the general public this whole number fraction style may be the best option but you will have to be careful about the form of words you use. For readers who are likely to understand a concise statement you might say:

'The probability that the next throw of two dice will score twelve is 1 in 36.'

But you could be more explicit:

'If you throw two dice many many times you will score twelve in 1 out of 36 throws. With fewer throws of the dice you may score twelve less often or more often than this.'

Probabilities depend on point of view

This is perhaps another surprising statement but easy to understand with a simple example. Imagine I toss a coin onto the table and cover it with my hand before you see it. I peek at the coin and ask you the probability of it being heads. From your point of view the answer is simple: 50%. I have a different point of view and a different answer: 100%. I can see the coin and it is definitely heads. A more down to earth example comes from the fact that, sometimes, medical decision makers have more rapid access to research data published in their own language: a doctor in Britain and a doctor in Russia may have different ideas of the probability of a treatment working because they read only journals published in their own language. Neither doctor is wrong in basing their decision on the information available to them, and a third person could only comment on which is making the better decision by having access to the research material that each is using to guide their judgement and comparing the quality of the science in them. Even then that person may arrive at a different answer because the treatment really works better (or worse) with Russians than with Britons.

Probabilities may be based on theory

Sometimes one is faced with a situation where the element of uncertainty is completely understood. For example, in the game of roulette efforts are made to ensure that there is an exactly equal chance of the ball falling in any of the numbered slots. That makes it straightforward to make calculations of probability. For example, if your selection is red, what is the probability that this will come up on the next spin of the wheel? There are 18 red slots and 37 slots in total so the probability is

$$\frac{18}{37} = 48.6486486486\%$$

It's possible to make this calculation just knowing how a roulette wheel is constructed and without doing any experiments or collecting any data. This type of figure is called an objective probability because the same result will be arrived at by anyone that works by the same model of the system in question. There are more serious scientific situations which can be dealt with in a similar way. For example, a physicist can calculate the probability that a molecule of gas in a container at a given temperature is moving within a given speed range. (The difficult maths was determined by Maxwell and Boltzman and the distribution of speeds is known as the Maxwell–Boltzman distribution.) Unfortunately, when it comes to medical tests and assays the reasons for uncertainty can rarely be analysed in this theoretical way and another approach is needed. You could view this as fortunate in a way because the mathematics behind objective probabilities is sometimes very complex. Objective probabilities are important in the fields of quantum mechanics, thermodynamics, various areas within structural molecular biology, spectroscopy and many other areas of physical science. They also feature in the theoretical study of random variation in measurements (more on this in later chapters).

Probabilities may be based on data

A probability based on data is called an empirical probability. Empirical means 'by experiment' or 'from experience'. Let's take a specific case. If you look at maternity ward records and count how many babies were born with jaundice you can estimate the probability that future babies will be born with jaundice. This probability is just a simple proportion. The probability of an outcome is the number of times this outcome was observed divided by the total number of observations. Florence Nightingale made this type of calculation to lend weight to her argument for improved hygiene in British army barracks. She calculated the probability of a soldier dying while living in army barracks and also the probability of adult males of the same age range dying while living at home. The mortality rate of men in barracks was much higher than the mortality rate of men living at home (presumably owing to unhygienic conditions). To underline this fact she then projected the increased strength, in terms of manpower, of the British Army if mortality rates in barracks were improved to match those of adult men living at home. This use of statistics and political lobbying to agitate for reform of the armed forces was Florence Nightingale's major achievement and so it is a shame that she is thought of by many as nothing more than the lady with the lamp.

An empirical probability is calculated by making a number of observations of a repeated event and counting the different outcomes. The assumption is that the observations in the past will be representative of events in the future. This doesn't mean that the past observations determine future events, only that the same (possibly unknown) factors that determined the past events will also determine future events. The great advantage of empirical probabilities is that they can be used in situations where there is no understanding of the mechanisms that lead to different outcomes. Let's take a rather abstract example.

Imagine I have an irregularly shaped rock that has nine, roughly flat facets. I've painted numbers from 1 to 9 on the faces and I ask you the probability of the rock landing on the ground on facet 1. Even if you examine the rock you won't be able to give a figure because the irregular shape will mean that some facets are more improbable than others. A physicist might measure the shape of the rock, assume an even density and make objective calculations of the probability, but let's imagine you don't have access to a physicist. The alternative is simply to throw the rock many times and count how many times it lands on facet 1. You don't need any understanding of physics and you don't need to make any examination of the rock to calculate a probability; you just divide the number of times it landed on facet 1 by the number of times you threw it:

$$P(A) = \frac{N_A}{N}$$

The probability of an outcome, $P(A)$, is the number of times this outcome was ✦ observed, N_A, divided by the total number of observations, N.

Probabilities depend on available information

It is quite possible for two people to arrive at two different probabilities for an event and for both answers to be perfectly valid. This may sound bizarre but it is an important principle. It helps to look at this separately for objective and empirical probabilities.

Objective probabilities

Maxwell and Boltzman worked out their distribution based on their understanding of the way that molecules move in a gas and interact with each other and the walls of the container. We can't be certain that their theory of gases is a perfect picture of how the universe works and it might be the case that another scientist working at the same period in history had a different, equally feasible theory which would lead to different probability figures. Until the point when the theories could be tested by experiment, both probabilities would have been equally valid. Therefore as theories develop on how systems work so the mathematics used to calculate probabilities will constantly be revised.

Empirical

Empirical probabilities involve taking a sample of events to predict events in the future. Inevitably, different scientists will take different samples of events and will arrive at different probability figures. The discrepancy in the case of empirical probabilities is because each figure is an estimate. Therefore you must be prepared to revise your probability if more data are collected.

Different situations

The probability that an asymptomatic person has colorectal cancer will be different in three different situations:

1. They have no test results.
2. They have a positive result on a faecal occult blood test.
3. They have a negative result on a faecal occult blood test.

Knowing the test result gives you a different probability. After all, there wouldn't be any point doing the test if that weren't the case.

Sampling

Imagine you are a doctor in general practice and want to know the proportion of patients on your register who are smokers. The information is not available in the medical records so you will have to run a survey. Consider these different strategies for the survey:

1. Telephone every patient on your records and ask them if they smoke regularly.
2. Select a small group of patients to be surveyed.

The first method will give you the most reliable figure because it involves surveying every individual in the group you are interested in. The alternative strategy is to select a sample from the population. (The word 'population' is used to describe the complete set of all individuals of interest even when it is just a small group such as the one in this study.) Clearly it will be quicker and easier to survey a sample of patients rather than all the patients on the register, but will the results be as good? The sample is taken to be representative of the entire population, and proportions are calculated on the sample data as if they were the population data themselves. This assumption breaks down if the sample is improperly selected from the population. Compare the following sampling strategies:

1. Get the receptionist to survey patients when they come to the clinic for an appointment for the next few weeks.

   This will bias your sample towards patients who frequently visit the clinic. You might have more smokers among this group than you would expect for all your patients.

2. Telephone every patient whose records are stored in the end filing cabinet.

   This might work, depending on how records are arranged in the filing cabinets. Perhaps your practice administrator has stored all the records for chronically ill patients in the end cabinet for easy access.

3. Telephone every patient whose National Insurance number has '1' as the last digit.

   This seems like a random method for selection, but can you be sure that the National Insurance number doesn't contain some coded information? Perhaps the number will relate to age in some way and this sampling method will tend to produce a sample with an atypical age profile.

4. Use your computer to randomly select 200 patients whom you will then telephone.

   This is the best method. Only if you knew in advance which patients were smokers, could you systematically select a sample that fits exactly with the population – but then you wouldn't need to do the survey.

Samples are selected from populations at random to ensure the best chance of that sample properly representing the population.

Calculations with probabilities

It's possible to calculate some probabilities from other probabilities. These mathematical operations are accompanied by logical operations within the statement in English that the probability figures are attached to. The mathematics is simple for anyone with a basic grasp of arithmetic but the use of logical statements in English is a dying art so many people find this type of work with probabilities challenging.

Notation

In very simple probability calculations it is common to use the symbol p to indicate a probability associated with a statement. In more complex problems you use a symbol to represent a statement and then use P() to indicate its probability. For example, H might represent the statement 'the next coin throw will come up heads'. With this notation P(H) indicates the probability that the next coin throw will come up heads.

This may not seem useful yet but it will avoid tangled pages of statements and calculations as things become more complex. For example, P(A AND NOT B) indicates the probability that statement A is true and statement B is not true.

The NOT operator

The simplest calculation to make using a probability is to use the NOT operator with a statement.

If the symbol A indicates that the next die throw will come up with a 6, the symbols NOT A indicate that the next die throw will *not* come up with a 6 (i.e. it will be 1, 2, 3, 4 or 5).

$$P(\text{NOT A}) = 1.0 - P(A)$$

So,

$$P(\text{NOT A}) = 1.0 - \frac{1}{6} = \frac{5}{6} = 0.8333333 = 83.33333\%$$

Combinations of independent probabilities – AND

Sometimes you are interested in the probability of combinations of events occurring. For example, if the probability of throwing a coin and getting heads $P(A) = 0.5$ and for throwing a die and getting six $P(B) = 0.1666667$, what is the probability of tossing a coin and throwing a die together and getting heads and 6?

When the two events are independent of each other (the outcome of one cannot influence the other) the calculation is simple: you just multiply the individual probabilities together. So:

$$P(A \text{ AND } B) = P(A) \times P(B)$$
$$= 0.5 \times 0.1666667$$
$$= 0.0833333$$
$$= 8.33333\%$$

You can think of this as a simple geometry problem. See Figure 1.1.

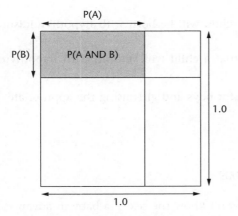

Figure 1.1 The probability P(A) is represented as a segment of a horizontal unit length. The probability P(B) is represented as a segment of a vertical unit length. P(A AND B) is simply the rectangular area within a unit square whose width and height are P(A) and P(B).

Combinations of dependent probabilities – AND

Sometimes when you look at combinations of events the probability of one event depends on the outcome of the other or vice versa. For example, you might suspect that the probability of a baby being born with hypothyroidism is different depending on whether the child is a boy or a girl.

Let's say:

A = child is born a girl

NOT A = child is born a boy

B = child is born with hypothyroidism

NOT B = child is born without hypothyroidism

You can get a very good estimate of P(A) and P(NOT A) because you have access to a very large number of records. Let's say that of 4 500 000 births in Britain, in recent years 2 280 000 were girls and 2 220 000 were boys.

$$P(A) = \frac{2\,280\,000}{4\,500\,000} = 0.50667$$

$$P(NOT\ A) = 1.0 - 0.50667 = 0.49333$$

A smaller number of records are available in which the incidence of hypothyroidism was recorded:

Of 100 000 girls, 22 were born with hypothyroidism.

Of 100 000 boys, 9 were born with hypothyroidism.

These can be used to calculate P(B) and P(NOT B). The problem is that we are admitting that P(B) could be different for boys and girls. How can we avoid getting mixed up between the two different probabilities?

A vertical line is the correct notation to use – it is short for 'given that'. Using the 'given that' notation:

P(B|A) = the probability that a child will be born with hypothyroidism given that it is a girl.

P(B|NOT A) = the probability that a child will be born with hypothyroidism given that it is a boy.

The calculation is done separately for boys and girls using the appropriate, separate data:

$$P(A\,|\,B) = \frac{22}{100\,000} = 0.00022$$

$$P(B\,|\,NOT\ A) = \frac{9}{100\,000} = 0.00009$$

Without an ultrasound scan you won't know the sex of a baby in advance. What if you want to know the probability that your next child will be a girl with hypothyroidism? When you calculate combinations with these dependent probabilities it's simply a matter of choosing the appropriate probability for the multiplication:

$$P(A \text{ AND } B) = P(A) \times P(B|A)$$
$$= 0.50667 \times 0.00022$$
$$= 0.00011147$$

This is the probability of a child being born a girl with hypothyroidism. If you want to know the probability that your next child will be a boy with hypothyroidism you also do a multiplication but you need to be careful to pick the correct two numbers:

$$P(\text{NOT } A \text{ AND } B) = P(A) \times P(B|\text{NOT } A)$$
$$= 0.49333 \times 0.00009$$
$$= 0.00004440$$

Geometrical representations are possible for these types of calculation also. See Figure 1.2.

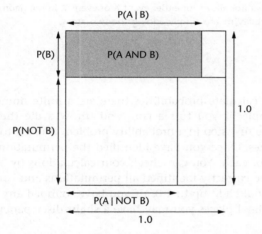

Figure 1.2 A geometrical representation is also possible when combining dependent probabilities but you have to be careful to choose the correct two probabilities.

Alternative events – OR

There is sometimes a need to consider two possible outcomes together or even a range of outcomes:

$$P(A \text{ OR } B) = P(A) + P(B)$$

The probability of scoring 1 on a die throw is $\frac{1}{6}$ and a score of 2 also has a probability of $\frac{1}{6}$. The probability of scoring 1 OR 2 is:

$$\frac{1}{6} + \frac{1}{6} = \frac{2}{6}$$
$$= \frac{1}{3}$$

A geometrical illustration of the OR operator is shown in Figure 1.3.

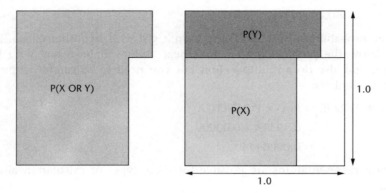

Figure 1.3 Using the OR logical operator allows for either event X or event Y. In the geometrical representation the combined probability means simply adding the two areas.

Permutations

With any situation where we calculate probabilities there are a finite number of different outcomes. For example, if you toss a coin and throw a die there are twelve different outcomes. The first step in a probability problem is often to identify all these different outcomes. Once you have identified the permutations you can calculate the probability of each. You can check your calculations by adding them all together – if you have correctly identified all permutations and correctly calculated the figures, they should add up to 1.0. (If you have rounded any numbers to a fixed number of decimal places you may have a slight discrepancy.)

Misconceptions about probability

There are many misconceptions about chance and about probability. People often come up against probability in three areas of their lives: accidents, health and gambling. It's inevitable that you will have to think about risk in your private life, but in employment too you may be responsible for assessing risk on behalf of others. You will need to communicate ideas to do with risk to people who may have all sorts of misconceptions about the subject. The following are some situations which illustrate various misconceptions that people have about probability.

Misconception – Small probabilities are zero

'The probability that you will die on the operating table is one in one thousand.'

'I'm not going to have this operation a thousand times so if I'm having it only once this means that I will definitely survive.'

Communicating to patients and their families the risk of treatment is very difficult and the above misconception is a common one. Here's how it might happen. 'I will have the operation only once so I need a something in one type of risk not a something in one thousand type of risk. I should scale the figures down by one thousand. If I divide one by one thousand I get more or less zero so the probability that fits me is zero in one.' It's unlikely that anyone consciously reasons it out this way but it is a basic problem with understanding small fractions.

'I don't understand. One thousand what?'

'Well, we did a study where we kept records of the success of the operation throughout the country. The operation was done 3026 times and three people died on the operating table.'

Sometimes it's simpler just to describe the study that led to the risk assessment.

Misconception – The probability must have been invalid because the unlikely event occurred

Some serious diseases are very rare but for one reason or another become highlighted in the news media. For example, the new variant of Creutzfeldt–Jakob disease, which is thought by some to be caused by eating beef from cattle suffering mad cow disease (bovine spongiform encephalopathy, BSE). This has caused serious concern among the British public. The following is the type of statement that might have been made by the father of a victim of this disease:

'They said getting this disease was only a one in ten million chance but my son is dead so they were wrong.'

This is a difficult misconception to pinpoint and certainly a misconception that will be very difficult to correct without offending this grieving parent. The problem is owing partly to the poor communication of a scientist's quantification of the risk of getting this disease but mostly because this father is confusing a probability that predicts whether people will get the disease in the future with his own experience of the past.

The original statement of risk might have been:

'We estimate that the probability of a randomly selected individual in the UK contracting the new variant of Creutzfeldt–Jakob next year due to contact with BSE infected beef is 0.0000001.'

In essence this scientist is predicting that from the population of the UK of about 60 million, roughly six will get the disease in the following year. The father in our case may have read this statement (or a version of it edited for a newspaper article) before his son fell ill and felt that he understood it. However, after the death of his son we might imagine the following dialogue:

'We stand by our risk assessment.'

'There is no doubt – my son is dead. How can you continue to say the risk was low?'

This father feels that there is a contradiction between the following two statements;

'They said it was extremely unlikely that my son would get the disease.'

'My son has died from the disease.'

There is no contradiction between the statements – no matter how low the risk, if the probability is greater than zero then an individual can get the disease. The main problem with the original probability estimate for this grieving parent is that it predicted how many would get Creutzfeldt–Jakob disease but didn't predict who would get it. When an individual has actually contracted Creutzfeldt–Jakob disease any probability is irrelevant – there is no uncertainty that they now have the disease.

Misconception – Past observations have a supernatural influence on future outcomes

Many people around the world gamble on lotteries. Most lotteries operate in a similar way: a few numbers are randomly selected from a pool of numbers. The participants get to choose any set of numbers they like and they win if their numbers come up in the draw. Lottery draws occur at regular intervals. It is very common for lottery gamblers to avoid selecting numbers that have been in recent winning draws. They believe that last week's numbers are less likely to come up this week than numbers that haven't come up before. What would be the physical mechanism that connects last week's draw with this week's draw? Perhaps the winning balls were made slightly greasy when they were handled last week and tend to stick to the sides of the drum – but this can't be true because they used a different set of balls this week. If you believe that the people who run the lottery are running it fairly then you might as well use the same numbers every week and pay no attention at all to previous draws. (You could exploit the misconceptions of your fellow gamblers by deliberately choosing numbers that have appeared recently. Then if you win you are less likely to have to share the prize with other gamblers.) There are many situations where past events can influence the probability of a future event, but only when there is a real physical mechanism that makes the connection. For example, if you have had heart/lung transplant then an influenza infection is more likely to be fatal in the future than it was before you had the operation. There is an easily understood reason for this: when one has an organ transplant you have to take drugs to prevent your immune system rejecting the organ and these also reduce your ability to combat viruses that enter your body.

Here's another gambling example. Imagine you have thrown a die and it has come up with a 6. You are about to throw the die again. Is a 6 more likely to come up on the second throw because 6 has just been thrown or is it less likely? Most people who are asked this question think that a 6 becomes less likely after a 6 because two sixes in a row should be unusual; others think that another 6 is more likely because the die has shown 'form'. Few people – those who have not studied probability – realize that the first throw cannot change the probability of 6 coming on the second throw. Those that get the answer wrong are demonstrating a mental ability that could aid survival in more natural circumstances – the ability to recognize patterns in their surroundings. The false hypothesis that the first throw can influence the second throw implies that the die can somehow remember its previous throw and from that influence the next throw. That doesn't make sense scientifically, and so in reality for the second throw a 6 is just as likely after a 6 as it is after any other number.

Misconception – What was the question?

The most difficult misconception results from an inability to ask the right question in clear logical English. This is particularly noticeable when combinations of probabilities are used, because that requires the careful use of logic words such as 'and', 'or' and 'not' to join together clauses in a statement. Another problem is the correct identification of the group being considered. The sentences below all use the same vocabulary but each has a distinct meaning and there will be a different calculation required for each.

'What proportion of the whole population will be diagnosed with colorectal cancer in the proposed screening programme?'

'What proportion of those receiving positive results on a faecal occult blood test will eventually be diagnosed with colorectal cancer in the proposed screening programme?'

'What proportion of people who have colorectal cancer will get a positive result from the faecal occult blood test recommended for the proposed screening programme?'

'What proportion of people who have colorectal cancer will not get a positive result from the faecal occult blood test recommended for the proposed screening programme?'

Try your own language comprehension skills with the scenario presented in the next section that relates to DNA fingerprinting.

Case study – DNA fingerprinting

You are the defence lawyer for a man accused of murder and you know that the police found that your client's DNA fingerprint matches skin found under the fingernail of the murder victim. The police took a blood sample from your client some years ago when he voluntarily participated in a quite separate investigation in another town. They arrested him for murder only after they compared the scene of crime sample with every DNA fingerprint in their records. The man has no known connection with the victim, no known motive for murder, no previous record of violence and there is no evidence other than the DNA fingerprint. A deposition by the manager of the forensic laboratory that runs the DNA fingerprints provides the following statement:

'The probability that an innocent person will have a matching DNA fingerprint is one in five million.'

It looks bad for your client, but ask yourself this question:

What is the probability that someone with a matching DNA fingerprint is innocent?

Is it one in five million? This is not a question of mathematics, it's a question of logical English. Are the two statements equivalent? It will help to tidy up the two statements by identifying their different components. There is an element to do with the skin sample belonging to the individual and an element to do with having a matching fingerprint:

Statement A: The skin sample came from the individual.

Statement B: The individual has a DNA fingerprint that matches the skin sample.

How does the statement from the forensic lab put these together? They are taking innocence as a given and in that situation asking how likely it is that the fingerprint will match. So if you reword it as follows:

What is the probability that an individual has a DNA fingerprint matching the skin sample GIVEN THAT the skin sample does NOT come from the individual?

In symbols:

P(B|NOT A)

Let's do the same analysis of the question you asked. You are taking the matching fingerprint as the given and then you ask for the probability of innocence. So, reworded it is:

What is the probability that the skin sample came from the individual GIVEN THAT they have a DNA fingerprint that matches the skin sample?

In symbols:

P(NOT A|B)

It's now clear that these two probabilities are different. So what is the answer to this second question? You ask the forensic expert and the answer comes back:

'It's about 50%.'

The first probability given by the forensic expert was useless for the purposes of your client's trial but it might easily have convinced a jury to convict because the impressively small probability distracts from the words in the statement in English associated with it. The main point of the statement was to say that almost all innocent people would be cleared if they were tested – a quite correct point but not very relevant to this trial. However, of the thousands of people who *could* have committed the crime there might be roughly two different people with the same DNA fingerprint. The police happened to have the DNA fingerprint of one of these two men on file and it is a one in two chance that he did it. When you present your client's case, how will you help the jury to avoid misunderstanding the figures? You need to phrase carefully the questions that you ask the forensic experts, but you may find that the jury is just not capable of telling the difference between the two different statements given above, no matter how good you are at explaining.

A case similar in some ways to the one above did happen in Britain some years ago. The individual was found guilty, partly because the judge instructed the jury that the phrase 'reasonable doubt' does not equate to a specific numerical probability and in effect stopped the jury from considering the statistical evidence. Current rules in Britain say that the police can collect DNA samples as part of the investigations for a crime but this evidence must not be kept on file and used for another, unrelated crime. This is intended specifically to prevent the type of problem presented above, but when an investigation involves DNA fingerprinting of the population of an entire town, miscarriages of justice are possible because the scientific evidence hinges on the estimation of probabilities.

Summary

- The most important point made in this chapter is that the science of probability involves an unbreakable link between mathematics and logical English. If you want to assess your own understanding of this point you should try Exercises 2, 3, 4, 9 and 10 in Chapter 9. You can also study Chapter 7 which is devoted to a detailed case study involving policy making based on data and probability calculations.

- Probability can be used reliably to communicate ideas involving uncertainty. See Exercises 4, 9 and 10 in Chapter 9.

- Probability can be used to make decisions. See Exercises 7 and 8 and Chapter 7.

- There are a variety of conventions for presenting probabilities. Test your ability to convert between these different conventions with Exercise 1.

- Empirical probabilities are estimated from repeated observations. See Exercises 2, 3, 7 and 8.

- Probability estimates depend on the available data, and researchers who collect their own data may obtain estimates that are different but equally valid. See Exercises 2 and 3.

- Logical statements can be combined using logic words and there are corresponding combinations of probability figures that use simple arithmetic. Exercises 6, 7 and 8 give you practice of these operations.

- The ability to assess risk based on day-to-day experience is a natural ability for humans and works well in many natural situations but we often fail when we attempt to apply this 'common sense' approach to modern-day situations. We can reach conclusions more reliably when we use data and quantitative methods but sometimes that means we have to work against our deeply ingrained misconceptions about probability.

Chapter 2

Significance

Introduction

In science we ask questions, propose answers to those questions and then design experiments that will test the validity of those answers. It is rare that experiments completely confirm or completely disprove our ideas and so the mathematics of probability is often applied to the results of experiments to quantify the likelihood that our hypothesis is true. A significance test is often involved in experiments where the result involves some uncertainty. Before we move on to a definition of significance, we need a suitable example to illustrate the explanation and we need to define a few terms used to describe scientific method.

In this chapter, a significance test called the χ^2 test (chi squared test) is presented. This is mathematically a fairly simple test to perform and the text concentrates on the general principles of significance testing, particularly the null hypothesis. Decision making in many situations, medical, scientific or industrial, is often based on the outcome of significance tests, and so the last part of the chapter discusses the consequences of taking action based on levels of significance.

Defining some terms

Theory

A theory is a explanation for a set of facts or phenomena. Theories can be created, starting from known facts and phenomena, purely by reasoning. Theories are not statements of fact or truth.

Hypothesis

A hypothesis is a clear statement which can be tested by experiment. The development of one or more hypotheses from a theory is a crucial part of scientific method. A hypothesis predicts the outcome of an experiment. When the experiment is done the hypothesis will be proved true or false depending on the result. Some hypotheses support a theory if they are true, and other hypotheses may work against the same theory if they are true. One of the best ways to test the strength of a theory is to try to break it, and so it is common to devise hypotheses that will contradict a theory.

Proof

The verb 'to prove' means to test. (When you prove bread you find out whether the yeast is alive or dead.) An unproved hypothesis is one for which no experiment has been made. Proof always aims at one of two outcomes: finding that the hypothesis is true or false. Sometimes 'proved' is used in the sense of 'proved and found true' and the word 'disproved' is used in the sense 'proved and found false'.

Horseradish and the faecal occult blood test

The basis of the faecal occult blood test mentioned in the previous chapter is the peroxidase activity of haem. If a tumour is bleeding into the bowel, haem will be present in the faeces and haem can be detected because it breaks down peroxides. Horseradish contains high concentrations of a peroxidase enzyme and, if present in the diet, it may cause false positive results on a test for blood in the faeces. (Actually, horseradish contains so much peroxidase that it's used as a raw material by manufacturers of laboratory supplies. Peroxidase is a useful component of a number of immunological assays.)

The theory

We can theorize:

If people eat food that contains substances which can also break down peroxide, these substances may survive the digestive system and they may produce false positive results on the faecal occult blood test.

This is a quite clear-cut theory that immediately suggests experiments we could do. Why not just go ahead and design the experiment? Well you could do that but you will have to decide after the event what specific hypothesis the experiment tested. It's better to devise the hypothesis first if you want to make sure that you're going to do the right experiment.

Hypothesis

The problem with the theory above is that it is little too general – it doesn't talk about specific foods or specific substances in food or a specific group of people for testing. It also says people 'may' get positive results, which is a bit vague. We need to be more specific with our hypothesis if we want to be able to test it. So let's propose:

A diet including horseradish sauce will cause an increase in the rate of positive results for a group of subjects who are free of colorectal cancer because horseradish sauce has peroxidase activity.

Horseradish has natural peroxidase activity so it is a prime candidate for causing false positive results according to our theory. We could come up with a long list of other hypotheses mentioning other foodstuffs with peroxidase activity (for example, turnip and melon) and if they all turn out to be proved true they will add support to the theory. To be thorough we ought to devise a number of hypotheses mentioning foodstuffs without peroxidase activity and these will support the theory if they are proved false.

Experiment

The hypothesis leads to a very obvious experiment.

1. First we need a group of people who are definitely free of colorectal cancer.
2. Next we randomly assign them to either the treatment group or a control group. The treatment group will have a diet including horseradish sauce and the control group will not. (There will be exactly the same number of people in each group – that isn't really important for the experiment but it will make the maths easier to follow later in this chapter.)
3. After being on the diet for one or two days the volunteers will provide a faeces sample.
4. The faecal occult blood test will be run on all the faeces samples and will be assigned positive and negative results.
5. The count of positive and negative results will be made separately for the treated group and the control group.

(We'll skip over some of the details of the experiment, like choosing amounts of horseradish in the diet and the choice of the other components of the volunteers' diet.)

Results – Frequencies

Here are some (fictitious) results from the experiment above. Each of the figures is a count of the number of individuals in that category. This type of data is called frequencies, and the table of frequencies, Table 2.1, is called a histogram. Sometimes the word 'histogram' is used of a bar graph that presents the results – see Figure 2.1.

Table 2.1

	Control	Treatment
Positive	15	21
Negative	123	117

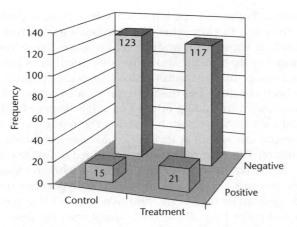

Figure 2.1   The histogram shows that there were a few more positives and a few less negatives in the treated group.

Have we proved that the hypothesis is true? There are certainly more positive results in the group that were on the horseradish diet. We ought to consider alternative explanations for the result before we assume that our original hypothesis is supported by the data. Are there other possible hypotheses? Let's list some next.

Alternative hypotheses and the null hypothesis

1. A diet including horseradish will cause an increase in the rate of positive results for a group of subjects who are free of colorectal cancer because it has peroxidase activity.

2. A diet including horseradish will cause an increase in the rate of positive results for a group of subjects who are free of colorectal cancer because it causes inflammation and bleeding in the bowel.

3. Horseradish sauce contains an ingredient, other than horseradish, which will cause an increase in the rate of positive results for a group of subjects who are free of colorectal cancer.

4. Someone in the lab wants to make sure the theory is supported and is prepared to cheat – for the tests he processed he put down a few extra positive results in the column for the treated group.

5. **Individuals produce false positive results on a faecal occult blood test for a number of unknown reasons and the ONLY reason that there are more positive results in the treated group is because the random method used to select volunteers for the treated and control groups accidentally put more of these people into the treated group.**

We now have a real problem. What we really wanted was a true or false proof of the original hypothesis but now we are starting to think of multiple explanations of the result. One of these hypotheses – the last – is a special one. Hypotheses 1, 2, 3 and 4 can happily coexist with each other, but if hypothesis 5 is true all the others must be false. Look at them again and see if you agree. There might be peroxidase activity in horseradish, it might also cause inflammation of the bowel, there may be another substance in horseradish sauce that causes positive results *and* someone may have been cheating, but the word 'only' in hypothesis 5 denies the truth of all other hypotheses (including others you might think of later).

Hypothesis 5 is special for another reason: it is saying the results came out the way they did due to no special mechanism at all. They could have just turned out the way they did by pure chance. This type of hypothesis is called a **null hypothesis**. No matter how many alternative hypotheses you come up with to explain the results of an experiment you can always work out the null hypothesis that contradicts all of them. The great advantage of the null hypothesis is that if you can prove it to be true you will prove that all the other hypotheses are false. So you are back to a true/false situation. Sadly, if the null hypothesis is proved false you are still left with several alternative hypotheses to distinguish among. We need some statistics at this point.

Significance

We now have a perfect situation for explaining the meaning of the word 'significance' because it is tied up with the null hypothesis. If the null hypothesis is proved false then the experimental results are said to be significant. If the null hypothesis is proved true then the experimental results are said to be not significant. How does this relate to the experimental data? If there was a very big increase in the number of positive results in the treated group then that would be a significant result, but if there was just a very small increase in the number of positive results in the treated group that would not be a significant result. Put another way, a significant result is one that supports the hypothesis that some real mechanism caused the observed result. But how big does the difference have to be if we are to call it significant, and just how small does the difference have to be if we are to call it not significant? The dilemma is solved by treating significance as a probability. It's a value on a sliding scale from 0.0 to 1.0 or from 0% to 100%. It is the probability that the null hypothesis is *not* true. The following section of this chapter looks at significance tests in general and then we will return to the data collected in this experiment and work step by step through an analysis of it.

Significance tests

Statisticians have provided us with a large range of different significance tests. These are like the choice of tools in a mechanic's toolbox because different mathematical methods suit different jobs. It's useful to look at what makes significance tests different and what they have in common.

Common features

1. There are some data to work with.
2. There is some random element that affects the results.
3. There is always a null hypothesis to prove.
4. The data are processed and the probability of the null hypothesis $P(H_0)$ is calculated.
5. The significance can be calculated $P(\text{NOT } H_0) = 1.0 - P(H_0)$.

Differences

Different mathematical methods are needed because experiments produce different types of data. Here are some examples of different types of data:

1. Data consist of counts in a two by two table.
2. Data consist of repeated measurements of a value.
3. Data consist of repeated measurements categorized into a two by two table.
4. Each datum is the measurement of two different properties of an individual.

We'll look at one of these situations now and the others in the following chapters.

The chi squared significance test

The appropriate significance test to apply to a table of frequencies that we have in the horseradish experiment is the chi squared (χ^2) test. (The symbol χ is the Greek letter chi, which is pronounced 'ky' to rhyme with sky.) If you have a computer program that has a χ^2 facility it is very easy to get a result. The basic steps that are involved are described below.

Step 1: Decide on a null hypothesis

Significance tests always concentrate on the null hypothesis. It is normal first to assume that the null hypothesis is true and then to find out how certain you are. We've already discussed the null hypothesis for our horseradish experiment. We are saying that the only reason there are more positives in the treatment group than the control group is because of random variation.

Step 2: Quantify variation

The null hypothesis mentions random variation. If we assume the null hypothesis is true, all the variation is random variation, but just *how much* variation do we have? It's important to quantify it. For our frequencies we will have to work out what values we would have expected if there were no variation at all. The measure of variation we need to calculate for our data is the statistical value χ^2 (which is why the test is called a chi squared test). A computer can be given the observed frequencies and the expected frequencies and will calculate χ^2 for you.

Step 3: Quantify significance of variation

This is the part that involves some serious statistical theory. However, it's the same statistical theory for every chi squared test no matter what the experiment involves, and that means you can trust the statisticians and the computer programmers and just get a computer to do the computation for you. The probability of the null hypothesis being true, $P(H_0)$, is a function of χ^2 and the degrees of freedom.

Step 4: Interpret results and report conclusions

This is a step that people often neglect. You really need to relate the result to your hypotheses and your theory and communicate these conclusions effectively to others.

Step 1 requires an understanding of the experiment, and the computer can't help you. Step 2 requires you to work out some expected values – the computer can't do that automatically because it isn't capable of understanding your null hypothesis and how it relates to the data. Actually calculating a χ^2 value from observed and expected frequencies is not difficult but the computer can calculate it for you more reliably. Step 3 can be done entirely by a computer because it just involves taking the χ^2 value and doing some hard maths on it to find a probability. Step 4 is down to you, but the computer can help with word processing the report.

Observed and expected frequencies

By observed frequencies we just mean the data from the experiment. It's useful to total up the rows and the columns as shown in Table 2.2

Table 2.2			
Observed	**Control**	**Treatment**	**Total**
Positive	15	21	36
Negative	123	117	240
Total	138	138	276

Now we need expected values. The biggest problem with understanding χ^2 is getting to grips with what is meant by expected values. The word 'expected' is shorthand for the more complex phrase 'The frequencies you would expect to see *if* the null hypothesis were true and random variation were eliminated'. What does that mean for our experiment? It won't make any difference to the fact that exactly 138 people were in the control group and exactly 138 people were in the treatment group, but it implies that the same number of positive and negative results are expected in the two groups. The total number of positive results was 36 and we would expect these to be evenly distributed between the control group and the treatment group. The total number of negative results was 240 and we would also expect these to be evenly distributed between the two groups. That allows us to complete an 'expected' results table – Table 2.3.

Table 2.3

Expected	Control	Treatment	Total
Positive	18	18	36
Negative	120	120	240
Total	138	138	276

You should notice that the totals are unchanged but we have balanced the positive and negative results in the two groups.

Calculating χ^2

If you have a suitable computer program you can simply type in Tables 2.2 and 2.3 (without the totals) and select the appropriate command or function. Actually the calculation is quite easy to do on a pocket calculator and the maths is explained at the end of this chapter. For now, let's assume you have a computer and it has provided the result:

$\chi^2 = 0.799$

The result doesn't have units because it is derived from frequencies. The more variation there is in the observed results the bigger χ^2 will be.

Calculating $P(H_0)$

Finding $P(H_0)$ from χ^2 involves mathematics which is complex and time consuming if you try to do it with an ordinary pocket calculator. There are two ways to deal with this. The traditional way is to use a table of statistical values. A χ^2 table will have a large number of χ^2 values printed against the corresponding probability values and you just find the χ^2 value that is closest to your value and read off the probability. The alternative is to use computer software to compute the probability. Actually, which ever method you use there is an extra bit of information needed apart from the value of χ^2 and that is called the **degrees of freedom (d.f.)**. For now, just accept that there is one degree of freedom for our experiment, but the subject is covered in the last part of this chapter in more detail.

For $\chi^2 = 0.799$ and d.f. $= 1$

$P(H_0) = 0.372$

$\qquad = 37.2\%$

You can also quantify significance by finding $P(NOT\ H_0)$:

$P(NOT\ H_0) = 1.0 - P(H_0)$

$\qquad\qquad = 1.0 - 0.372$

$\qquad\qquad = 0.628$

$\qquad\qquad = 62.8\%$

That's the end of the calculations and we're ready to interpret the result.

Interpretation

We now have the ability to make some statements about the results of the experiment. All we need to do is take the null hypothesis and reword it to include the probability figure:

There is a 37.2% probability that the ONLY reason that there are more positive results in the treated group is because the random method used to select volunteers for the treated and control groups accidentally put more of these people into the treated group.

You can add a little human judgement to that statement and say:

The experiment did not show a very significant effect (62.8%) and does not provide conclusive support for the theory about peroxidase activity of elements of diet and their effect on the faecal occult blood test.

What would you say if the probability $P(NOT\ H_0)$ was not 37.2% but was 5% or 50% or 90% or 99.99%? At what point would you say that the results really support the original theory?

Decision making

We have looked at how scientific theories lead to hypotheses and how these lead to experimentation. It is rare that this chain of events ends there though. In pure research the results of experimentation lead to development of theory, more hypotheses and more experimentation, where the direction taken will at each stage be affected by the data obtained and the outcome of analysis of those data. Medical-related research is not just concerned with finding out how the human organism works but it also aims to find ways to prevent and cure disease, and so there are often crucial decisions about health policy to be made based on experimentation.

What health policy decisions might arise from our faecal occult blood test example? Obviously we would want to minimize false positive results to avoid causing distress to healthy members of the public and we may advise people to avoid eating certain foods before providing a faeces sample. We don't want to stop people eating horseradish sauce if there is no need to do so. So there is a simple binary (only two options) decision to make: either advise people to avoid horseradish before they have a faecal occult blood test or don't advise people to avoid horseradish. We have a binary decision to make and we have a significance level to base it on. If the significance were 99.9999% we would certainly recommend the new advice, and if the significance were 0.0001% we would not; but what about intermediate figures? Clearly we need to select a specific significance level and take action depending on whether the experimental significance came above or below that threshold – but what threshold?

Critical probability

A threshold value used with a significance test is called a **critical probability**. It is a probability that you choose to help you make decisions based on experiments that end in a significance test. You must recognize that from time to time you will make the wrong decision in these situations. In the following explanation we will stick with the horseradish issue, so let's summarize the link between experimental results and decision making:

Significant result: $P(H_0)$ is small = take action

Not significant result: $P(H_0)$ is big = don't take action

If we give the critical probability the symbol p_{crit} we can say:

If $P(H_0) < p_{crit}$ then take action

If $P(H_0) > p_{crit}$ then DON'T take action

What are the consequences of getting the decision wrong? There are two different ways to get it wrong: taking action when action wasn't needed, and not taking action when action was needed. There are special statistical terms for these two types of mistake. Sadly the word 'error' is used, but this has no connection with the use of the word 'error' as defined in the next chapter. This special use of the word is always linked with the words 'type I' or 'type II' so this will help you avoid confusion.

Type I error
You make a type I error (I is roman numeral one) when you incorrectly reject the null hypothesis. You will take action when you shouldn't have taken action.

Type II error
You make a type II error when you incorrectly accept the null hypothesis. You will not take action when you should have taken action.

Your policy on choosing a critical probability depends on how serious the consequences of making a type I error are, balanced against how serious the consequences

LIVERPOOL
JOHN MOORES UNIVERSITY
AVRIL ROBARTS LRC

of making a type II error are. This is where mathematics starts to give way to judgement and morality. Let's stick with the mathematics a bit longer. Our experiment looked only at horseradish in the diet but the original theory should lead us to take a closer look at a whole range of foodstuffs. So we will conduct a series of experiments and at the end of each we will have to decide whether to take action with respect to that foodstuff. For every foodstuff we test we will have the opportunity to do the wrong thing, and the number of type I and type II errors will depend on the choice of critical probability. (We won't actually know how many type I and type II errors we have made no matter what critical probability we choose.)

Low critical probability
What will happen if we choose a low value for our critical probability? What we are saying is that we will easily accept the null hypothesis even when the results are quite significant. That means that we will not take action in relation to many foodstuffs and that means we will tend to make many type II errors and few type I errors.

High critical probability
What will happen if we choose a high value for our critical probability? What we are saying is that we will easily reject the null hypothesis even when the results are not very significant. That means that we will take action in relation to many foodstuffs and that means we will tend to make many type I errors and few type II errors.

A critical probability of $0.05 = 5\%$ is a conventional choice for situations where the consequences of type I errors are not too serious. For our example, unnecessarily advising people not to eat a specific food for a few days is not a serious fault, so a 5% critical probability is suitable. If taking action was potentially harmful to the public's health we might choose a 1% or even a 0.1% critical probability, but then we would have to consider the harm done by type II errors. In our case a type II error means that many more people might suffer the distress of a false positive result from the faecal occult blood test because they are eating foods that affect the results. There are really no easy answers here but we benefit from the systematic approach that the statistical method provides us with.

The mathematics of χ^2

You may sometime be forced to calculate χ^2 values with a pocket calculator and this last section of the chapter explains how to do the calculations by hand. Another reason for studying this section is to provide you with a way to check that your computer software is doing what you think it should do.

Degrees of freedom

Before we get in to the calculations we need to explain degrees of freedom. This is a tricky idea to get your head around and we will benefit from a general explanation and an explanation that is specific to the chi squared test.

The degrees of freedom are the number of data points you need to know before you can deduce the remaining data points from them and the statistical value in question.

This is a bit of an obscure statement and it helps to make this more specific to chi squared tests.

The degrees of freedom are the number of entries in the observed frequency table you need to know before you can deduce the remaining frequencies from the column and row totals.

Table 2.4 shows our data with only one frequency filled in.

Table 2.4			
Expected	**Control**	**Treatment**	**Total**
Positive	15		36
Negative			240
Total	138	138	276

You only have to do a few subtractions to find out what the three missing values are from the one data point given, and the same would be true no matter which of the four data points were given. So for this experiment there is one degree of freedom. The general rule for a table of frequencies is:

d.f. = (Columns − 1) × (Rows − 1)

There are some special-case chi squared tests where the expected values are based not just on the column and row totals but also on some hypothetical parameter, and in these special cases there may be a different number of degrees of freedom.

Small frequencies

There is an element of approximation in the statistical science behind chi squared tests which is usually inconsequential, but if any of the frequencies observed falls below 5 the approximation becomes much more serious and the significance test is no longer valid. Consequently, chi squared tests should not be applied if any of the frequencies is less than 5.

Calculating deviations

The first step in the calculation is to subtract the expected frequency E from the observed frequency O for each entry in the table. Actually we are not interested in whether the values are negative or positive, so what we want is the absolute difference. The absolute difference is sometimes denoted with vertical line brackets: $|O − E|$.

Table 2.5		
$|O-E|$	Control	Treatment
Positive	3	3
Negative	3	3

Table 2.5 is showing you how far the observed frequencies deviate from what you would expect if the null hypothesis were true and there was no random variation in the frequencies.

Continuity correction

It is at this point that we have another problem with an approximation. The statistical science behind χ^2 is based on numbers on a sliding scale but frequencies are in fact always whole numbers. This artificially increases the chances of type I errors and so some compensation is required. This is not a fiddle factor because it must always be applied when there is one degree of freedom and is not applied when there is more than one degree of freedom. The continuity correction is simple to make: you just subtract 0.5 from every frequency deviation, as shown in Table 2.6.

Table 2.6		
$|O-E|-0.5$	Control	Treatment
Positive	2.5	2.5
Negative	2.5	2.5

Calculating χ^2

When there is no need for continuity correction:

$$\chi^2 = \sum \frac{(O-E)^2}{E}$$

With continuity correction:

$$\chi^2 = \sum \frac{(|O-E|-0.5)^2}{E}$$

The capital sigma indicates that you make the calculation for each entry in the table and then add them all up. So:

$$\chi^2 = \frac{2.5^2}{18} + \frac{2.5^2}{18} + \frac{2.5^2}{120} + \frac{2.5^2}{120}$$
$$= 0.799$$

Calculating $P(H_0)$

The calculation of $P(H_0)$ depends on χ^2 and the number of degrees of freedom. The calculation is iterative: it involves the calculation of the product of a large number of individual terms. Consequently it isn't feasible to calculate it yourself with an ordinary pocket calculator. Therefore either you look up the nearest χ^2 value in a table of values or you use a χ^2 function in a computer program. If you were using Microsoft Excel, with χ^2 in cell A1 and the degrees freedom in cell B1 you could use the following formula:

=CHIDIST(A1, B1)

If you were using a book of concise statistical tables you wouldn't actually find the exact value of χ^2 but if you've chosen a fairly conventional critical probability (for example 0.05) you can find the critical χ^2 value in the table. You find the row in the table corresponding to a degree freedom of 1 and then run across to the column headed 0.05 to read off the critical χ^2 value.

If $\chi^2 > \chi^2_{\text{crit}}$ then it follows that $P(H_0) < p_{\text{crit}}$

If $\chi^2 < \chi^2_{\text{crit}}$ then it follows that $P(H_0) > p_{\text{crit}}$

Summary

- The word 'significance' is commonly used in a qualitative sense outside of science, but in a scientific context significance is a quantitative property.

- An estimation of significance always indicates the significance of a hypothesis. The sequence is usually: to devise a hypothesis, design an experiment to test that hypothesis and then to analyse the data from the experiment to determine the level of significance. Test your understanding of this by trying Exercise 11 in Chapter 9.

- Statistical estimations of significance are called significance tests and they rely on the identification of the correct null hypothesis for the experiment that was performed.

- The chi squared test is a significance test that can be applied to data in the form of frequencies. See Example 5 in Chapter 8 for a computer-based worked example. Exercises 12, 13 and 14 in Chapter 9 provide an opportunity to apply the chi squared test using a pocket calculator and with a computer.

- Often, binary decisions are based on probabilities, which are numbers on a sliding scale. That means that a critical probability must be selected. It also leads inevitably to type I and type II errors.

- The balance between type I and type II errors can be influenced by the choice of critical probability.

Chapter 3

Measurements and assays

Introduction

This chapter is about making measurements. In an ideal world this would be a very short chapter because what we would like to do is just make one measurement and that's that. Unfortunately, measurements of biological values are often complicated and involve some calculations. Also there is unreliability of one sort or another in every measuring system and you will have to do extra work to minimize it. When unreliability can't be completely eliminated you will always have to be well informed about the type and amount of the unreliability in the measurements. The trickier statistics for dealing with the unreliability of measurements will be left for the next two chapters, and in this chapter we'll stick to studying the types of unreliability and the reasons they occur. First we need to cover the basic vocabulary of this area of data handling.

Scale

Everyone is familiar with the ideas that different scales can be used to measure the same thing because we often see temperatures in weather reports quoted in centigrade and Fahrenheit. A scale defines a starting point and the distance between values. Both the starting points and distances between values are different with the centigrade and Fahrenheit scales. Another quality of scale is linearity. Most physical measurements like mass, length and so on are linear scales. In other words they relate in a linear way to the underlying system. For example, if you compare a 1 kg iron bar with a 2 kg iron bar, the mass of one is twice that of the other because it has twice the number of iron atoms in it. An example of a non-linear scale is pH. The acidity of a solution is dependent on the concentration of protons in it but there is not a linear relationship between pH and proton concentration. In fact each drop in pH of one unit represents a tenfold increase in proton concentration. This is called a logarithmic scale. Non-linear scales are very useful in chemical and biological systems and a section of this chapter is devoted to them.

Unit

It is vital that the proper units are quoted alongside measurements. There are internationally agreed units for physical measuring systems called Système International (SI). These allow variation on each of the basic units. For example, the basic unit of metres has variations such as kilometre, micrometre and nanometre. It is possible to makes mistakes with units that end with values one thousand times too big or one thousand times too small by using the wrong variation of the basic unit.

Assay

Historically the word 'assay' was used by metalworkers for measurements that indirectly indicated the purity of metals. Now the word is used more generally to indicate any kind of indirect measurement, especially measurements of the concentration or activity of biomolecules. Measuring a piece of wood with a ruler would not normally be called an assay, but measuring the concentration of glucose in a urine sample using a colour change chemical reaction would be called an assay. The colour intensity can be measured and that indirectly assays the glucose.

Error

This has a finely defined meaning in statistics: it is the distance of a number from the correct value. Usually it refers to the distance of a single measurement from the correct value. On occasions the word is used in combination with other terms and might indicate the distance between two numbers or an average distance between several numbers and a reference point. An alternative word for error would be deviation. In statistics this word rarely means mistake (the exception is in specific phrases such as 'type I error') and never refers to faults in the procedure by which the measurements were obtained. The word derives from the Latin verb *errare*, which means to wander.

Human error

This is a misleading term that is best avoided. It is used by some teachers of statistics to make the distinction between variation in numbers due to unreliability in the measuring system and real variation between different things being measured. It's misleading because it gives the impression that if people made the measurements correctly they would be perfect, but there is always some doubt about assay results no matter how competently they are made.

Estimate

An estimate is an attempt to determine a value. The word 'estimate' is used as a reminder that the number obtained is imperfect. An estimate may be a single measurement, the mean of several measurements or a number calculated from some other type of data.

Accuracy

This is a type of error. In practice you estimate accuracy by calculating the distance of your estimate from the correct value. However, it's a bit more complicated than that and will require a more mathematical explanation later in this chapter. If you write reports involving statistics you should avoid using this word in any non-statistical sense because it could cause confusion.

Precision

The most important point to make about precision is that it is not the same as accuracy, although it is another type of error. In the most general terms it is the width of the range of values, either side of your estimate, within which you believe the true value lies. This will become clearer as you study the mathematics. Like the term 'accuracy' it is best to avoid using the word 'precision' for non-statistical uses. The word 'exact' has no particular statistical meaning and so it's a useful alternative to the words 'accurate' or 'precise'.

Granularity

The granularity of a measuring system is the smallest distance between two values that the measuring system can produce. For example, a 30 cm ruler usually has a granularity of 1 mm because it has 1 mm divisions marked on it. A digital thermometer may display values to one decimal place and so has a granularity of 0.1 °C. Sometimes the word 'precision' is used in situations where granularity would be more appropriate.

Two measuring systems

In the discussions of scale, units and error it will be useful to consider actual measuring systems. In the biomolecular sciences, making measurements is rarely as easy as using a ruler or a top pan balance. With complex assays it can become quite a problem to trace all the possible sources of imprecision or inaccuracy. In the sections that follow, ideas will be illustrated using two measuring systems, one a simple physical measurement and the other a biochemical assay.

How long is a piece of string?

The simplest possible type of measurement is to use an ordinary ruler to measure the length of a piece of string. Measuring string will be used as a very simple measuring system to help illustrate some of the statistics.

How concentrated is a solution of lactate dehydrogenase?

Measuring string is not going to be a major activity for bioscientists and so this assay of an enzyme will be used as a second example to show how the statistics work with a real measuring system. The amount of lactate dehydrogenase in a patient's blood serum can be clinically useful in certain circumstances. The enzyme is found in all types of cells and appears in the blood serum because cells die and break open. Some diseases result in cell damage that leads to elevated levels of lactate dehydrogenase in the blood serum: heart disease and hepatitis for example. Consequently levels of lactate dehydrogenase can be monitored to check the progress of a disease or the success of a treatment. Some drugs, such as chemotherapy drugs for cancer, cause cell damage as a side effect and might also raise lactate dehydrogenase in the serum.

The serum lactate dehydrogenase assay

What if you wanted to determine the concentration of the enzyme lactate dehydrogenase in a solution? A direct way (i.e. a measurement as opposed to an indirect method, which would be called an assay) might be to use purification techniques to remove all other material, desiccate to remove the water and then weigh the residue; but this just isn't going to work. The problem with a direct measurement is that it's impossible to purify the enzyme perfectly – some of the enzyme will be lost and some contaminants will remain. Inaccuracy and imprecision will both be large. The same argument is true for proteins in general and also for other biomolecules. An assay is indirect because it measures some property that indicates the concentration of the biomolecule.

Lactate dehydrogenase catalyses the oxidation (dehydrogenation) of lactate using NAD^+ as oxidizing agent. Actually, the reverse reaction is favoured – the reduction of pyruvate by NADH:

$$\text{Pyruvate} + \text{NADH} + H^+ \rightleftharpoons \text{Lactate} + NAD^+$$

The concentration of NADH in a solution can be measured in a spectrophotometer because it absorbs ultraviolet light at a wavelength of 340 nm. So the assay for the enzyme depends on finding the rate of the reaction that the enzyme catalyses by following the consumption of NADH. Concentrations for the various reagents are chosen to give the optimum rate of reaction for the enzyme and so that small errors in those concentrations will have negligible effect on the rate.

The assay is carried out in the following way:

1. The blood sample is spun in a centrifuge and serum is taken from the supernatant.

2. 0.05 ml blood serum and 2.5 ml of 0.2 mmol l^{-1} NADH are mixed in a cuvette. (A cuvette is a square container that is 1 cm across and transparent to ultraviolet light.) The NADH is buffered to a pH of 7.2.

3. The cuvette is placed in the spectrophotometer. The cuvette holder will have a thermostatically controlled heat jacket set to 30 °C. Time is allowed for the mixture to warm up to 30 °C.

4. 0.5 ml of 1.6 mmol l⁻¹ pyruvate is mixed into the cuvette.

5. A chart recorder creates a paper trace that indicates the consumption of NADH over the next 3 minutes.

6. The amount, in moles of NADH consumed is calculated from the drop in absorbance.

7. The rate of consumption of NADH is calculated.

8. The rate of consumption of NADH per litre of blood serum is calculated.

The units of the assay are micromoles per second per litre, μmol s⁻¹ l⁻¹ (sometimes referred to as μkatals per litre).

You may not be happy with the units arrived at in this assay because they aren't in the form of mass per unit volume. However, they are perfectly adequate for clinical purposes. Doctors are interested in comparing a patient's value with values from normal patients, patients with heart disease, patients with hepatitis, and so on, so they don't really care what the units are, just so long as they are on a consistent and reliable scale.

Units

The unit of length, the metre, is easy to understand. There is a metal bar devised in France that defines one metre in length. Any object that has the same length as the standard is also 1 m. Because the scale is linear, a length of 2 m is simply double the length of the standard metre. You will be familiar with the convention for following a measurement with a symbol representing units of the value. There are two issues relating to units that you may not be familiar with. First, the correct way to present very large and very small values, and secondly, a way to check your calculations using units. Both of these require a close look at the way units work in algebra. Let's take a simple example:

240 km

We can treat the symbols 'k' and 'm' as variables in algebra. In fact the k (short for kilo) does stand for the number 1000 so we can say:

240 km = 240 × 1000 × one metre

Multiplying by the symbol for one metre may seem a little strange because a metre isn't a number. However, a unit can participate in algebra just like any variable.

Big and small numbers

There are two strategies for dealing with numbers that are very big or very small: using prefix symbols or using powers of ten. The international system (SI) for units allows the use of prefix symbols that increase or decrease the size of the unit in powers of ten and these are in such widespread use that they are impossible to ignore, but many scientists prefer to state explicitly a power of ten using numbers. Table 3.1 shows a selection of big and small lengths.

Table 3.1		
Unmodified number	Prefix	Power of ten
1000 m	1 km (kilometre)	1×10^3 m
0.01 m	1 cm (centimetre)	1×10^{-2} m
0.001 m	1 mm (millimetre)	1×10^{-3} m
0.000001 m	1 μm (micrometre)	1×10^{-6} m

Checking your calculations with dimensional analysis

With simple measurements like length it's difficult to make a mistake with the units, but with complex assays it's easier to make a mistake in the calculations or a mistake writing the correct units next to the number that you calculate. Let's run through the calculation you need to make to find the number of moles of NADH consumed in a run of the lactate dehydrogenase protocol described earlier. What you have is the light absorbance at the start of the reaction and again after the reaction has proceeded for a period of time. There are also figures needed in the calculation that are the constants for the protocol. A_1 and A_2 are the absorbance values (variables with no units). Let's imagine that you've run the experiment with a test solution:

$A_1 = 1.020$ $A_2 = 0.980$

There are some numbers determined by the experiment:

Width of cuvette = 1.0 cm

Reaction volume = 3.05 ml

If you look in a data handbook you will find:

Molar absorption coefficient of NADH = 6.230 ml μmol^{-1} cm^{-1}

(In English this is saying that if you made up a 1 micromole per millilitre solution of NADH and put it in a 1 cm cuvette the absorbance would be 6.230. The absorbance would be proportionately lower if the concentration were lower or if the path of light through the cuvette were lower.)

Now we have all the numbers we need, but what do we do with them? Do we multiply by the molar absorption coefficient or divide? You are probably quite confident about how to do the calculation but maybe not totally confident. Dimensional analysis allows you to check whether the calculation you intend to do is consistent with the units you expect. The key to dimensional analysis is to make sure that you always include the units when you write out your calculations.

Let's say that the following calculation is the one you propose to make:

$$\frac{(1.02 - 0.98) \times 3.05}{6.230 \times 1.0}$$

The problem with writing out the calculation like that is that it contains no clues about whether you've got it right. The key to dimensional analysis is to make sure you always include the units when you write out your calculations. Here is the same thing with the units included:

$$\frac{(1.02 - 0.98) \times 3.05 \text{ ml}}{6.230 \text{ ml } \mu\text{mol}^{-1} \text{ cm}^{-1} \times 1.0 \text{ cm}}$$

You can highlight the fact that you are doing algebra with the units by including the multiplication symbols with the units:

$$\frac{(1.02 - 0.98) \times 3.05 \times \text{ml}}{6.230 \times \text{ml} \times \mu\text{mol}^{-1} \times \text{cm}^{-1} \times 1.0 \times \text{cm}}$$

The next step is to separate the units from the numbers by rearranging. It's just a matter of changing the order of the multiplications:

$$\frac{(1.02 - 0.98) \times 3.05 \times \text{ml}}{6.230 \times 1.0 \times \text{ml} \times \mu\text{mol}^{-1} \times \text{cm}^{-1} \times \text{cm}}$$

It's easier now to see that some of the units cancel each other out. To start with, the ml that appears above the division line will cancel the ml that appears below it. So:

$$\frac{(1.02 - 0.98) \times 3.05}{6.230 \times 1.0 \times \mu\text{mol}^{-1} \times \text{cm}^{-1} \times \text{cm}}$$

Now you can cancel the cm and the cm^{-1}. (Note that cm^{-1} is the same as 1/cm so $\text{cm} \times \text{cm}^{-1}$ is the same as cm/cm.)

$$\frac{(1.02 - 0.98) \times 3.05}{6.230 \times 1.0 \times \mu\text{mol}^{-1}}$$

It would be useful to get the remaining unit out of the division term:

$$\frac{(1.02 - 0.98) \times 3.05}{6.230 \times 1.0} \times \frac{1}{\mu\text{mol}^{-1}}$$

The remaining unit is an inverse of an inverse, i.e. $(\mu\text{mol}^{-1})^{-1}$, which gives you μmol as your final unit.

$$\frac{(1.02 - 0.98) \times 3.05}{6.230 \times 1.0} \times \mu\text{mol}$$

You've now determined the units, so all that remains is to do the arithmetic, thus:

$$= 0.019582665 \ \mu\text{mol}$$

The units match the ones we expected. You can be more confident about the numerical part of the calculation because the units came out right. Most of the mistakes you could have made with the original calculation would have resulted in the wrong units. Let's take an example. Imagine you multiplied by the molar absorption coefficient

$$\frac{(1.02 - 0.98) \times 6.230 \text{ ml } \mu\text{mol}^{-1} \text{ cm}^{-1} \times 3.05 \text{ ml}}{1.0 \text{ cm}}$$

This calculation will result in the incorrect answer:

$$= 0.76006 \ \mu\text{mol}^{-1} \text{ ml}^2$$

These units are obviously crazy but you might have believed the number you calculated if you hadn't included the units. It's the units that show that something has gone wrong, and that's the benefit of dimensional analysis. When you talk about slip-ups in the calculation of assay results it is important that you avoid the word 'error' because that refers to a statistical idea. The word 'mistake' is much better for this type of thing. For the sake of completeness, the final assay result involves finding the rate of consumption of NADH per litre of serum in the reaction. Including the units again, the correct final answer is:

$$\frac{0.01958266 \; \mu mol}{180 \; s \times 0.00005 \; litre} = 2.1758516 \; \mu mol \; s^{-1} \; l^{-1}$$

The final value is within the normal range for an adult. Phew!

It's worth saying a little now about the number of decimal places given in the calculated answer, 2.175851614. The number has nine decimal places and ten significant figures and that is probably giving a false impression of the reliability of this assay. Later in this chapter the granularity of the lactate dehydrogenase assay is discussed and you will find advice on how to choose the appropriate number of significant figures.

Calibration curves

On occasions one might need to know a concentration of an enzyme in mass per unit volume. The answer to this is to assay a series of standard solutions of lactate dehydrogenase that have been prepared with measured quantities of purified protein. This is only done if the extra effort is justifiable, because units of $\mu mol \; s^{-1} \; l^{-1}$ are for some reason unacceptable. From the assay results a calibration curve is graphed. Now, when you assay a sample you take the reaction velocity and use the calibration curve to look up the concentration. (Calibration curves are often straight lines, but any mathematician will tell you that straight lines are just a special type of curve.) Figure 3.1 shows a typical calibration curve.

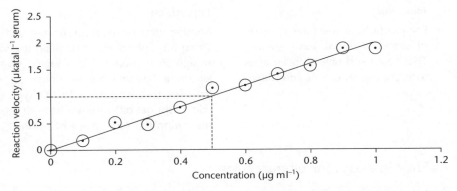

Figure 3.1 This calibration curve for enzyme X was produced by assaying dilutions of a standard solution. A serum sample gave 1.0 $\mu katal \; l^{-1}$ and the calibration curve shows that this corresponds to a concentration of 0.5 $\mu g \; ml^{-1}$.

Statistical error

If you want to understand error properly then you need to take a mathematical approach. Error in a single measurement can be shown mathematically as:

$$X = x + E$$

This equation shows that a measured value, X equals the true value, x plus an error E. When you assay biomolecules you need to be concerned with detecting and quantifying error. If you're going to do that you need to distinguish between the two types of error: inaccuracy and imprecision.

Inaccuracy	*Imprecision*
Is a fixed positive or negative error that is the same if the measurement is repeated.	*Is a random positive or negative error that varies every time the measurement is made.*

The total error is:

$$E = F + R$$

So here is a modification of the previous equation: F is fixed error and R is random error. Hence:

$$X = x + F + R$$

Knowing the maths is not the whole picture – how does the error occur?

Error in a simple measurement

Simple physical measurements such as using a 300 mm ruler to measure the length of a piece of string are subject to error:

Inaccuracy	*Imprecision*
For example, if the ruler is made of wood it might have shrunk. That could add several millimetres to every measurement you make.	Another error would result from stretching the string too tight or not stretching it enough to straighten it properly. This would give you a length a few millimetres longer or shorter than the true length. With this type of error you might get different results from repeatedly measuring the same piece of string.

Error in assays for biomolecules

What about error in assays? It can be trickier to spot sources of error in assays than in simple physical measurements. Take the serum lactate dehydrogenase assay. Some possible sources of error are listed below.

Inaccuracy

1. The red cells can be damaged by rough handling of the blood sample and if ruptured will release more lactate dehydrogenase into the serum. This will affect all the measurements done on the sample and will lead to inaccuracy.

2. NADH can be chemically degraded to various products when stored for many days. Even very small amounts of degradation products are a problem because the altered molecules can bind competitively or non-competitively with the enzyme and stop the NADH binding. This inhibition will slow down the reaction in every measurement you make. This will lead to inaccuracy.

3. Lactate dehydrogenase is degraded. If the sample to be measured is stored at very low temperature it is damaged and many of the molecules of enzyme stop working. Samples are slowly degraded over a period of days even when stored at the ideal temperature of 4 °C. Degraded enzyme will lead to a lower reaction velocity for every measurement. This will lead to inaccuracy.

4. Lactate dehydrogenase standard is degraded. If the purified enzyme used to create the calibration curve is degraded then the calibration curve will be wrong. All of the assay results based on this calibration curve will give concentrations that are consistently high. This will lead to inaccuracy.

Imprecision

1. The automatic pipette is being used in fluctuating temperatures. It is important for the assay that exactly the same volumes of reagent and sample are mixed in the cuvette each time the assay is run. Liquid is drawn into an automatic pipette below a volume of air; it doesn't matter what temperature the air is at, *but* if the temperature changes after you start to draw liquid then problems occur. If the pipette gets warmer as you draw liquid then the air in the chamber expands and the volume of liquid taken in will be less than the correct value. If the pipette cools as you draw liquid then the air in the chamber shrinks and too much liquid is taken. In normal circumstances, temperature fluctuations cause random variation and hence imprecision, but some bad practices will favour only warming up errors and some inaccuracy will accompany the imprecision.

2. The temperature of the cuvette varies. The assay depends on a constant temperature of 30 °C. Small variations won't make much difference to the rate, but if the thermostat is poor and the temperature slowly fluctuates between 25 °C and 35 °C some measurements will have faster or slower reaction rates as a result. This will lead to imprecision.

With symbols

So the first result from an assay X_1 will deviate from the true value x owing to two types of error: random R_1 and fixed F. Either error component might be positive or negative.

$$X_1 = x + F + R_1$$

If a second assay is made, the true value x must be the same and, by definition, so will the fixed error F but the random error will be different:

$$X_2 = x + F + R_2$$

The same reasoning applies to all subsequent measurements. Here the random error in a measurement is represented by a single symbol but imprecision actually consists of many random events occurring in an assay protocol. You might say:

$$R_1 = r_{1,2} + r_{1,3} + r_{1,4} + r_{1,5} + \ldots$$

Replicate measurements

When you make a measurement you would like to do it just once and be able to rely on the figure you get, but as we have seen, error may result in a value that deviates from the true value. With just one measurement there is no way to tell whether there is error and no way to distinguish between inaccuracy and imprecision. Fortunately there is a straightforward way to deal with imprecision. (We will return to inaccuracy later.) If you repeat your imprecise and inaccurate measurement of a piece of string you will get different values. Because imprecision is due to positive and negative random errors, when measurements are repeated you start to see a range of values which you can expect will lie either side of x. Consequently you get a better idea of what x is by repeating the measurements. If measurements of a single value are repeated without alteration to the method they are called **replicate measurements**. Figure 3.2 illustrates the point with a sporting analogy.

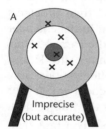
A
Imprecise
(but accurate)
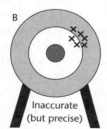
B
Inaccurate
(but precise)
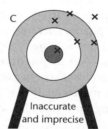
C
Inaccurate
and imprecise
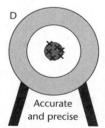
D
Accurate
and precise

Figure 3.2 Accurate and precise archery. When you have replicate measurements, the distinction between accuracy and precision becomes clearer. Fred is an archer and is trying out four new bows. Bow A is poorly weighted and is difficult to keep steady. Bow B has sights that are misaligned. Bow C has both faults but bow D is nearly perfect. Fred shoots six arrows with each bow, and the results are shown here. With only a single arrow you would be unable to tell the difference between imprecision and inaccuracy, but the difference becomes clear when six arrows are shot. The same is true of replicate measurements.

Estimates from replicate measurements

You have to apply some simple statistics to the replicate measurements to obtain a single estimate of the value and also to estimate the precision that was achieved.

The data from a set of replicate measurements vary only because of the imprecision (by definition the error due to inaccuracy is constant).

Mean

You need to take this set of numbers and find a single number that best represents the true value being measured. This statistic is called the **mean**. The mean is a balance point in the middle of the data set and is represented with a bar: $\bar{X}$. For example, the mean of 1, 3 and 8 is 4.

$$\bar{X} = \frac{\Sigma X}{N}$$

It's quite straightforward to calculate a mean for N measurements. The capital sigma indicates the sum of the measurements. N is the number measurements that were made. Hence:

$$\Sigma X = X_1 + X_2 + X_3 + \ldots$$

Since the only thing that varies from measurement to measurement is the random error you can see that the mean measurement is equal to the true value plus a fixed error plus the mean of the random errors:

$$\bar{X} = x + F + \bar{R}$$

It's likely, but not certain, that the mean will be closer to $x + F$ than any single measurement that you might pick out. The more measurements you make the closer $\bar{R}$ will get to zero and the closer the mean will get to $x + F$.

So $\bar{X}$ is an estimate of $x + F$
The point that is most important to note is that $\bar{X}$ is not an estimate of x unless you know for certain that your assay is perfectly accurate.

Precision

If you make a number of measurements and they differ from each other then you must have imprecision – but how much imprecision? It is just as important to estimate precision as to calculate the mean. To estimate precision you always calculate the standard deviation of the replicates and you may also calculate the standard error and confidence limits. These are all subjects of the following chapter. However, we will deal with the special case when all the replicate measurements are exactly the same numbers.

Granularity versus precision

Let's imagine that you measure the length of a metal rod using a ruler with millimetre marks on it. The end of the rod lies between the 220 mm and 221 mm marks on the ruler. You offer the rod up to the ruler repeatedly and every time you get the same result. Does that mean the measurements are perfectly accurate and precise? No, the results may be both inaccurate and imprecise. Let's take accuracy first. There is no way to know whether the ruler we are using has not warped or shrunk. If it has then the results will be inaccurate, and since this is a fixed error it will not matter how many times we repeat the measurement, we won't detect the problem. To find out for certain we would have to compare the ruler with a standard length.

With every measurement agreeing, one might be tempted to say that the precision is perfect; but what would be the effect if there was a very small amount of imprecision? If the random elements in each measurement were around ±0.1 mm they would not cause any distinction between the measurements because the ruler has 1 mm marks on it. So we have imprecision but we can't detect it. This is because the granularity of the measuring system is much greater than its precision. In measuring systems, where the smallest unit is bigger than the random variation, precision becomes irrelevant and granularity is what determines the upper and lower limits on the true value. You might make a statement like this: 'My estimate of the length of the rod is 220.5 mm. Assuming that my ruler is accurate (I haven't checked it), I am confident that the actual value is between 220 and 221 mm.'

In this type of situation it is common to hear the word 'precision' when the word 'granularity' would be more correct.

Granularity of the lactate dehydrogenase assay

The granularity of ruler measurements is obvious, but what is the granularity of the lactate dehydrogenase assay? It's a puzzle because the assay starts with readings from a spectrophotometer but those figures are manipulated mathematically before you get a final answer. To start with, the spectrophotometer has granularity; often 0.001 but sometimes smaller. To find the granularity of the assay, just put 0.001 into the same calculations you would apply to a change in absorbance when you run the assay:

$$\frac{(0.001) \times 3.05}{6.230} \ \mu\text{mol} = 0.000489567 \ \mu\text{mol of NADH consumed}$$

Expressed as rate per litre serum:

$$\frac{0.00048957}{180 \text{ s} \times 0.00005} = 0.05439629 \ \mu\text{mol s}^{-1}\text{l}^{-1}$$

Since the absorbance readings are always whole number multiples of 0.001, the assay will produce values that are always whole number multiples of 0.0544 μmol s^{-1} l^{-1}. Put another way, this assay protocol can't detect the difference between two serum samples if they are closer than 0.0544 μmol s^{-1} l^{-1} apart.

Take another look at the example assay result calculated earlier:

2.175851614 μmol s^{-1} l^{-1}

The number of decimal places given suggests a granularity of 0.000000001 µmol s^{-1} l^{-1}, which is plainly far better than we could ever achieve with a real assay. Since the granularity isn't a simple power of ten, there isn't an exact number of significant figures that matches the granularity.

2.2 µmol s^{-1} l^{-1} suggests granularity of 0.1 µmol s^{-1} l^{-1}

2.18 µmol s^{-1} l^{-1} suggests granularity of 0.01 µmol s^{-1} l^{-1}

The second of these would be the better choice. This advice applies because in this example the precision of the assay is unknown and the granularity is the only measure we have of the reliability of the results. If the estimate of precision was much worse than granularity we might reduce the number of significant figures even more. Could the granularity be improved by modifying the assay protocol? Table 3.2 looks at how the granularity was calculated and assesses the possibility of improving granularity.

Table 3.2

Calculation	Result	Possible improvement
Granularity of spectrophotometer	0.001	Perhaps you can obtain a better spectrophotometer
Volume of reaction mixture	3.05 ml	You can't reduce this number – the cuvettes are a standard size and a reduced volume would mean the light beam might not pass through the solution
Molar absorption coefficient of NADH	6.230 ml µmol^{-1} cm^{-1}	This is a physical property of NADH and can't be changed
Time between absorbance readings	180 s	This could be increased and would improve the granularity, *but* if you let the reaction continue too long you may find that the lactate produced starts to give a reverse reaction and the results will not be accurate
Volume of serum added	0.05 ml	This could be increased and would appear to improve granularity, *but* there may be increased imprecision because small random errors in the assay would be amplified

In most cases there is little point trying to improve granularity because imprecision is a much greater cause of error and because in some cases efforts to improve granularity may make the imprecision worse.

Logarithmic scales

In chemical and biological systems there are numerous situations where measurements on a linear scale are awkward to use. The acidity of solutions, the loudness of sounds and the growth rates of bacterial cultures are all things that are better measured on a logarithmic scale. Logarithms used to be much more widely understood before pocket calculators were invented but now it's possible for someone with quite good qualifications in mathematics from school to have very little experience with logarithms. The algebra of exponents and logarithms is presented here in terms of the more basic algebra with which it is assumed you are familiar.

Exponents

The best way to approach logarithms is via the more familiar subject of exponents, so it's worth reviewing a little algebra. You will be familiar with the notation of exponents like:

$$a^2 = a \times a$$

Every time the exponent goes up by one you multiply by a, and every time the exponent goes down by one you divide by a. Here are ten observations about exponents.

1. The zero power of any number is one

$$a^0 = 1$$

You can see the sense of this if you start from a^2 and reduce the exponent by one and then by one again:

$$a^2 = 1 \times a \times a \qquad a^1 = 1 \times a \qquad a^0 = 1$$

2. Negative powers are fractional values

$$a^{-1} = \frac{1}{a}$$

This follows naturally on from the previous point. If you keep subtracting one from the exponent, you will be dividing by a until you end up with a fractional value:

$$a^2 = 1 \times a \times a \qquad a^1 = 1 \times a \qquad a^0 = 1 \qquad a^{-1} = \frac{1}{a} \qquad a^{-2} = \frac{1}{(a \times a)}$$

It's common to see a^{-1} notation to indicate a reciprocal instead of $1/a$.

3. Multiplying a^m by a^n

$$a^m a^n = a^{(m+n)}$$

This makes sense if you write out the exponents as a series of multiplications:

$$a^3 a^2 = a \times a \times a \quad \times \quad a \times a = a^5$$

This is a very useful feature of exponents. If the base values are the same you can perform a multiplication by just adding the exponents.

4. Rearranging fractional exponents

$$\frac{1}{a^n} = a^{-n}$$

If you take the inverse of a value with an exponent you can just negate the exponent.

5. Division

You can put together the previous two principles to help with division:

$$\frac{a^m}{a^n} = a^{(m-n)}$$

Put another way:

$$\frac{a^m}{a^n} = a^m \times \frac{1}{a^n}$$

$$= a^m a^{-n}$$

$$= a^{(m-n)}$$

6. Powers of powers

$$(a^m)^n = a^{mn}$$

For example:

$$(a^2)^3 = a^2 a^2 a^2$$

Applying point 3 gives:

$$a^{(2+2+2)} = a^{(2\times3)}$$

$$= a^6$$

7. Fractional powers (roots)

Exponents need not be whole numbers and there is a special root symbol for certain fractional powers:

$$a^{1/n} = \sqrt[n]{a}$$

Finding the nth root is the reverse function of finding the nth power because:

$$\sqrt[n]{a^n} = (a^n)^{1/n}$$
$$= a^{n/n}$$
$$= a^1 = a$$

8. More fractional powers

It follows from the previous point that:

$$a^{m/n} = \sqrt[n]{a^m}$$

9. Two bases raised to the same power

This and the next point deal with two different bases raised to exponents so we have another symbol, b, in the equations:

$$a^n b^n = (ab)^n$$

For example:

$$a^2 b^2 = aabb$$
$$= abab$$
$$= (ab)^2$$

10. Two bases in a division

It follows from the previous point that:

$$a^n/b^n = (a/b)^n$$

There are a number of exercises on exponents at the end of this book with which to test yourself.

What are logarithms?

You are already familiar with the use of exponents to avoid many zeros in the notation for very large or very small numbers. For example:

$$520\,000 = 5.2 \times 10^5$$

or

$$0.0000034 = 3.4 \times 10^{-6}$$

The first part of the number can always be brought into the range 1 to 10 (or −1 to −10) by choosing the appropriate integer exponent of ten. From a mathematical point of view there is an added complication when you use exponent notation because you end up with two separate parts to the number you want to represent. Logarithms go a step further than the standard exponent notation because they are fractional exponents, and that allows you to bring the first part of the number to exactly 1.0. The following example will help.

Start with 520 000. We can find two integer exponents that give a value above 1 and a value below 1.

$$520\,000 = 5.2 \times 10^5$$
$$= 0.52 \times 10^6$$

So there must be a fractional exponent n between 5 and 6 that would allow you to write:

$$520\,000 = 1.0 \times 10^n$$

In fact $n = 5.716$ and so:

$$520\,000 = 10^{5.716}$$

There is a special notation for finding the logarithms of numbers:

$$\log_{10} 520\,000 = 5.716$$

The subscript 10 is there to remind you that this is a base 10 logarithm.

Why use logarithms?

There are measuring systems that naturally produce a number which is already the logarithm of some underlying value that isn't directly measured. Take a look at this list of noises and think how you would arrange them on a scale from 0 to 10 of loudness (relative to the loudest set at 10):

Distant rustle of leaves   Low voice   Vacuum cleaner   Train   Small aircraft

It's likely that you would evenly distribute these on the scale in the order they are printed, and an electronic sound meter would more or less agree with your ears. However, the underlying physical system would disagree. The loudness is determined by the distance of the in and out displacement of your ear drum. Compare the different types of sound in Table 3.3.

Table 3.3

Sound	Relative perceived loudness	Displacement of ear drum	$\log_{10}$ of displacement
Distant rustle of leaves	2	0.1 nm	−1
Low voice	4	1.0 nm	0
Vacuum cleaner	6	10.0 nm	1
Train	8	100.0 nm	2
Small aircraft	10	1000.0 nm	3

Perceived or measured loudness is a natural logarithmic scale because each increase in perceived loudness is actually caused by the ear drum displacement being multiplied. If you just want to communicate the loudness of a sound to others, the logarithmic scale doesn't have any particular consequences. But if you want to make calculations for physical or electronic systems that relate to sound, you may well have to work with the algebra of logarithms. One mathematical consequence is that different logarithmic scales may choose different reference points for the value zero and may have different bases (10, 2 and the mathematical constant e are the most common bases). You will need to understand the algebra of logarithms to be able to convert between different logarithmic scales.

Units of logarithms

Do logarithms have units? You can answer the question yourself: take a look at Table 3.3 and imagine the displacement column was given in μm instead of nm. What would happen to the logarithm column? A 1000-fold increase in the figures in the displacement column will simply add log(1000) to the log column. So no matter what you do to scale up or down the original numbers, the logs will stay on the same scale but will be offset up or down. So logarithms don't have units but they do have a reference point that defines where the log, 0.0, will lie. They also have a base, 10 in the loudness example above. Loudness of sound is always measured by reference to a sound of standard loudness that defines the zero point for the logarithmic scale. Negative values indicate sounds that are quieter than the reference and positive values indicate sounds that are louder than the reference. On a log scale, silence would have to be indicated with a symbol for minus infinity, $-\infty$.

Algebra of logarithms

So far we've looked only at base 10 logarithms because they relate to the scientific notation for numbers that you are familiar with. However, it is possible to work with logarithms for different bases. For example, with base 2 any number can be represented:

$x = 2^n$ where $n = \log_2 x$

We can use the symbol b to represent any base and then:

$x = b^n$ where $n = \log_b x$

Here are five rules of algebra related to logarithms.

1. Log of 1

No matter what the base is, the log of 1 is 0:

$\log_b 1 = 0$

2. Log of the base

The log of the base number is always 1:

$\log_b b = 1$

3. Addition of logs

Adding the logs of two numbers is the same as the log of their product:

$$\log_b x + \log_b y = \log_b (xy)$$

4. Multiplying logs

Multiplying the log of a number is the same as the log of its power. For example:

$$3 \log_b x = \log_b (x^3)$$

5. Converting between bases

You can easily convert the log of a number in one base to the log in another base by multiplying by a constant:

$$\log_c x = \log_b x \times \log_c b$$

For example, to convert between base 10 and base 2:

$$\log_2 x = \log_{10} x \times \log_2 10$$

pH

Measuring the acidity of a solution with pH is an obvious example of a logarithmic scale. There are several reasons for using a logarithmic scale:

1. The meters for measuring acidity produce a voltage that is related to the logarithm of the proton concentration.
2. The scale is more useful because many of the calculations you want to make involve the logarithm of the proton concentration anyway. This is because a number of principles of chemistry involve exponential relationships (including the principles that govern the way pH electrodes work).
3. Some biological systems are sensitive to acidity and you get much more regular shaped graphs if you put pH on the x-axis rather than proton concentration.

Like linear scales, logarithmic scales need a zero point, and where linear scales need magnitude a logarithmic scale needs a base. With pH the reference point is a solution of 1.0 moles per litre of protons and the base is 10. So we have a simple scale that relates to the underlying concentration of protons in the solution. Most solutions you measure are less acidic than the reference point and that would make most logs negative, so for simplicity with pH, unlike with most other log scales, the log is multiplied by −1:

$$pH = -\log_{10}[H^+] \quad \text{with } [H^+] \text{ measured in mol l}^{-1}$$

Some example proton concentrations and their corresponding pH are shown in Table 3.4.

Table 3.4	
Proton concentration, [H⁺]	**pH**
1.0 mol l^{-1}	0
0.1 mol l^{-1}	1
0.001 mol l^{-1}	3
0.0000001 mol l^{-1}	7
1.0×10^{-14} mol l^{-1}	14

Wider use of logarithms

Logarithms are particularly of use when you study the relationships between variables for biological systems. An exponential relationship is quite common. For example, growth rate of bacterial cultures is exponential while the nutrient broth is in excess and it can be useful to compare the log of dry mass of bacteria against time instead of directly comparing dry mass and time.

Summary

● Assay results are estimates because they are subject to error.

● Error takes two different forms, imprecision and inaccuracy. Exercise 18 in Chapter 9 is a problem-solving exercise that tests your understanding of the difference between these two forms of error. You will need to be familiar with the statistical values from Chapters 4 and 5 to complete the exercise.

● The raw data from an assay protocol, for example, readings from a spectrophotometer, often need to be mathematically processed to obtain an estimate in the proper units. Exercise 15 in Chapter 9 will test your ability to use a spreadsheet to implement the appropriate calculations for a lactate dehydrogenase assay.

● Mistakes can be made in the mathematical processing of raw data from an assay and dimensional analysis can be used to check those calculations. For practice of this technique see Exercise 16.

● Assays often make use of a calibration curve to convert raw data or partially processed data from an assay into the desired units. Exercise 17 provides practice of constructing and using a calibration curve.

● Sometimes assay results are very small or very large numbers and calculations involving assay results rely on the proper use of scientific notation. See Exercise 19.

● Processing assay results may involve the use of exponents or logarithms. For practice of the appropriate algebra try Exercises 20 and 21.

Chapter 4

Precision

Introduction

The precision of an assay has a critical effect on its clinical, scientific or industrial usefulness. Over the past few decades the science of immunology has produced numerous precise assays for biomolecules where previously there was no practical way to measure these substances. It's important to realize that this advance in the precision of assays has nothing to do with improvement in the quality of laboratory staff – that is the myth of human error at work. It is developments in the protocols for running assays that has given us these improvements in precision. Each new protocol reduces the effect of random variations on the final result. (If you are unsure about the meaning of precision go back to study the previous chapter.) The statistics we shall study are the normal distribution, the t distribution, standard deviation, standard error and confidence limits. But first, let's look at a case study that highlights the importance of precision.

Case Study Precision of the α-foetoprotein assay

There are various techniques for visualizing the different proteins in a blood serum sample. One way is 2D polyacrylamide gel electrophoresis (2D PAGE). 2D PAGE gives us a rectangular slab of gel with spots of protein stained blue. There is generally the same pattern of spots for blood samples from different people, but in 1956 it was noticed that foetal blood gave a pattern with a prominent extra spot in the region of the gel called α-1. This newly discovered protein was named α-foetoprotein. Given sufficient blood plasma, it's possible to purify α-foetoprotein. If the protein had some enzymatic activity it would be possible to devise an assay for it in the same way that the lactate dehydrogenase assay was developed, but unfortunately no enzymatic activity was found. Consequently it was not possible to create an assay for α-foetoprotein when it was first discovered and there was no obvious clinical advantage to knowing its concentration at that time anyway.

Around 1965 a new class of assay was developed for proteins that works with many proteins, not just those with enzymatic activity. These are called immunoassays because they exploit immunological reactions. The basic idea is to produce antibodies specific to the protein to be assayed by injecting a solution of the purified protein into an animal and then collecting blood serum from it after an immune response has been provoked. Antibodies are a type of protein that bind to a recognized target molecule and are part of an organism's defence mechanism against foreign bodies such as micro-organisms. When antibodies to α-foetoprotein come into contact with the α-foetoprotein from a serum sample they bind with it. The assay developed in the 1960s involved dropping blood plasma into a well cut into a gel slab containing antibodies. As the blood plasma diffuses into the gel, a ring of cloudy precipitated protein forms. The more α-foetoprotein in the serum, the bigger the cloudy ring will be. Concentrations of the protein in foetal blood vary during foetal development; 2000 μg ml^{-1} would be typical.

Assays for α-foetoprotein have changed for the better since the 1960s but they still rely on immunological principles and involve measurement of how much antibody binds with the protein. The first immunoassays were very imprecise – the random variation affecting each measurement meant that you could get widely differing values for repeated measurements of the same blood serum sample. An improved version of the original assay involved forcing the serum proteins through the gel using an electrical field. Although still imprecise the new assay for α-foetoprotein uncovered some surprising facts: α-foetoprotein is present in the amniotic fluid that surrounds a developing foetus at a concentration 100 times lower than in the foetal blood, typically 20 μg ml^{-1}.

Higher levels of α-foetoprotein were found in the amniotic fluid if the foetus had a neural tube disorder, and so the small improvement in the precision of the assay opened up the possibility of diagnosing anencephalopathy and types of spina bifida early in pregnancy. (In a normal foetus some protein enters the amniotic fluid via urine but with these disorders extra protein leaks out into the amniotic fluid from exposed tissue.) This discovery created a great incentive to study α-foetoprotein more carefully and a more precise assay was needed. A new development of immunassays in the late 1960s was the radioimmunoassay. This works on the same basic principle as earlier immunoassays, but the use of radioactive antibodies enables a more precise assay protocol to be designed. With radioimmunoassay, researchers could detect much smaller concentrations with a precision as much as one thousand times better than with previous methods. It was then discovered that α-foetoprotein naturally 'leaks' into the maternal blood from the foetus, so that pregnant women have concentrations of α-foetoprotein at about 0.02 μg ml^{-1}. A precise assay made it possible to predict disorders in the foetus using a simple blood test on the mother.

At the same time that foetal blood serum, amniotic fluid and maternal blood serum were under scrutiny it was also discovered that male and female adults with primary liver cancer had much elevated levels of α-foetoprotein. With a precise assay doctors were able to chart the levels of blood α-foetoprotein during the progress of disease in patients with primary liver cell cancer.

The clinical importance of precision

With pregnant women the research described in the case study showed that the graph of α-foetoprotein concentration in the mother's blood was similar for all women with normal pregnancies but there was a noticeable difference when the foetus is more likely to have a deformity. This means that a simple blood test on a pregnant woman can be used for screening. With the new precise assay, researchers discovered that all children and adults have some α-foetoprotein in their blood but normally at very low concentrations. The concentration of α-foetoprotein in the blood starts to rise at the earliest stages of primary liver cell cancer before symptoms start to appear and drop again if treatment is successful, so a graph of concentration against time can plot the progress of the disease. You can see that crucial clinical decision making depends on the reliability of numbers that come out of the assay for α-foetoprotein, and that means especially the precision of the assay. Assays for α-foetoprotein are probably as good now as they will ever need to be and there is not so much incentive for further improving precision. However, there may well be other biomolecules whose importance is not yet understood and which will benefit from new and more precise assay protocols.

Imagine you are a doctor starting work at a new hospital and with your first case you need an α-foetoprotein assay on a blood sample. The lab in your hospital provides you with a number: 10 ng ml^{-1}. How do you know that you can rely on that number and base a clinical decision on it? As things stand you don't know if it's a reliable number. It would be a little better if you at least knew what assay protocol your lab is using and then you could find out from the literature what the precision of that assay protocol should be if it's performed well. It would be much better if your lab told you the precision of its own assay protocol along with its estimate of the concentration. One way to do that would be:

10 ng ml^{-1} (95% confident that the true value lies between 9 and 11 ng ml^{-1})

Unless you quantify the precision you don't know what the number is worth. Compare these:

10 ng ml^{-1} (95% confident that the true value lies between 5 and 15 ng ml^{-1})

10 ng ml^{-1} (95% confident that the true value lies between 9.9 and 10.1 ng ml^{-1})

10 ng ml^{-1} (95% confident that the true value lies between 9.99 and 10.01 ng ml^{-1})

Why do experienced medics and scientists often ignore the precision of their assays? The problem is that many assays are extremely precise and we become complacent. If an assay protocol is precise way beyond what you need, any laboratory can achieve a performance that is adequate for your purposes. In such a circumstance you can neglect quantifying the precision and get away with it. The danger is that you take the same attitude with all the assays you perform and with the less precise assays you could get into real problems. In this chapter we discuss precision in terms of the two measuring systems introduced in the previous chapter: using a ruler to measure string and the lactate dehydrogenase assay.

The binomial distribution

This section on the binomial distribution is a short diversion from the topic of precision. The aim is to introduce the idea of probability distributions in general with the simplest possible example, the binomial distribution, before moving on to look at the normal distribution and the *t* distribution.

Suppose you take a one pound coin, an Irish punt, an Italian hundred lire, a French franc, a Barbados dollar and a Chinese one jiao and throw them all in the air. What is the probability of getting one coin coming up heads? Or the probability of getting two heads, or three heads etc.? This type of problem is called binomial because the final results are dependent on the total effect of a number of binary outcomes. A binary event is one that can turn out in just one of two ways, i.e. heads or tails. The first step to solving this problem is to look at all the possible outcomes. Table 4.1 shows all the permutations possible and, in the right-hand column, the total number of heads for each combination.

Table 4.1						
£	EI£	L	F	$	Jiao	Heads
T	T	T	T	T	T	0
T	T	T	T	T	H	1
T	T	T	T	H	T	1
T	T	T	H	T	T	1
T	T	H	T	T	T	1
T	H	T	T	T	T	1
H	T	T	T	T	T	1
T	T	T	T	H	H	2
T	T	T	H	T	H	2
T	T	T	H	H	T	2
T	T	H	T	T	H	2
T	T	H	T	H	T	2
T	T	H	H	T	T	2
T	H	T	T	T	H	2
T	H	T	T	H	T	2
T	H	T	H	T	T	2
T	H	H	T	T	T	2
H	T	T	T	T	H	2
H	T	T	T	H	T	2
H	T	T	H	T	T	2
H	T	H	T	T	T	2
H	H	T	T	T	T	2
T	T	T	H	H	H	3
T	T	H	T	H	H	3
T	T	H	H	T	H	3
T	T	H	H	H	T	3
T	H	T	T	H	H	3
T	H	T	H	T	H	3
T	H	T	H	H	T	3
T	H	H	T	T	H	3

T	H	H	T	H	T	3
T	H	H	H	T	T	3
H	T	T	T	H	H	3
H	T	T	H	T	H	3
H	T	T	H	H	T	3
H	T	H	T	T	H	3
H	T	H	T	H	T	3
H	T	H	H	T	T	3
H	H	T	T	T	H	3
H	H	T	T	H	T	3
H	H	T	H	T	T	3
H	H	H	T	T	T	3
T	T	H	H	H	H	4
T	H	T	H	H	H	4
T	H	H	T	H	H	4
T	H	H	H	T	H	4
T	H	H	H	H	T	4
H	T	T	H	H	H	4
H	T	H	T	H	H	4
H	T	H	H	T	H	4
H	T	H	H	H	T	4
H	H	T	T	H	H	4
H	H	T	H	T	H	4
H	H	T	H	H	T	4
H	H	H	T	T	H	4
H	H	H	T	H	T	4
H	H	H	H	T	T	4
T	H	H	H	H	H	4
H	T	H	H	H	H	5
H	H	T	H	H	H	5
H	H	H	T	H	H	5
H	H	H	H	T	H	5
H	H	H	H	H	T	5
H	H	H	H	H	H	6

There are 64 different combinations of heads and tails when you throw six coins because $2^6 = 64$. Each of the combinations is equally likely because for all the coins there is an equal chance of getting heads and for getting tails. Calculating the probability of getting various different numbers of heads out of six means calculating a probability distribution. The calculations are quite easy: the probability of getting an individual combination is $\frac{1}{64}$ and to get the probability of a certain number of heads just multiply $\frac{1}{64}$ by the number of different combinations that give the right total. Table 4.2 has the answers.

Table 4.2		
Heads	Probability (Fraction)	Probability (Decimal)
0	1/64	0.015625
1	6/64	0.09375
2	15/64	0.234375
3	20/64	0.3125
4	15/64	0.234375
5	6/64	0.09375
6	1/64	0.015625

With probability distributions it's often helpful to graph the probabilities. See Figure 4.1.

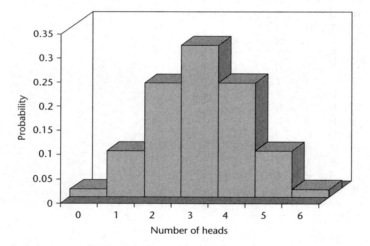

Figure 4.1 Bionomial distribution of six coin throws – a typical binomial distribution.

We won't spend any more time looking at the mathematics, but there are some observations to make:

1. The middle number of heads, three, is the most probable total.

2. A total of zero heads is very unlikely.

3. A total of six heads is also very unlikely.

4. A total of four heads is almost as likely as three heads. The same is true of two heads.

What happens when you actually try this out for real? Figure 4.2 shows the results of tossing the coins 250 times and counting the number of heads each time.

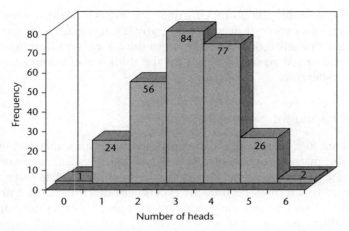

Figure 4.2 Binomial distribution of six coin throws. The observed frequencies would vary if the experiment were repeated.

You can see that the shape is roughly the same as the probability distribution but there is some variation. With a random event like coin tossing nothing is certain so there is no reason to expect an exact correspondence between the frequencies and the probabilities.

The normal distribution

Normal distributions often occur when random events affect repeated measurements of the same thing, for example replicate assay measurements. Therefore understanding imprecision means you need to understand the normal distribution. The normal distribution describes the frequency of different sized random errors in the measurement. Here is the equation from the previous chapter that describes error:

$$X = x + F + R$$

where x = the value to be measured

$\quad X$ = a measurement

$\quad F$ = fixed error

$\quad R$ = random error

The normal distribution describes the probability of obtaining a value of R within any given range. If measurements are repeated an infinite number of times the normal distribution will predict exactly the proportion of measurements that show a specific range of values for R. (In a similar way, although you must be uncertain that a single coin throw will come up heads, you can be more certain that from 1000 coin throws roughly 500 will come up heads.) If you study statistics as a subject in its own right you would need to study the complex mathematics that describes the normal distribution, but as a scientist who simply wants to use it as a tool you just need to know why it's important and how to use it. This is like the difference between learning to drive a car and learning how to design a car. You do need an overview of how the thing works but not a detailed theoretical understanding.

An imprecision thought experiment

Instead of working in a laboratory on an experiment you can sometimes imagine a system and use mathematics to predict the outcome of an experiment. Such thought experiments can be useful for helping us to conceptualize systems that are difficult to study directly. The thought experiment described in the next few pages will help you understand the origins of the normal distribution and subsequent explanations for the statistics will refer to it. This thought experiment will help you to understand the shape of the distribution, without the need for difficult mathematics. By following the instructions you can create five data points of your own. Creating your data will involve tossing a coin a large number of times. If you don't have the patience to do that you can study the method and then look at the sample results provided. Imagine that you want to measure the length of a very long piece of string using an ordinary 0.3 m ruler. You will need to measure out a number of 0.3 m lengths like a merchant who measures out cloth against his arm. Each 0.3 m length will be subject to random error and these errors will accumulate as you work along the string. You can make the mathematics of this system quite easy to understand. For example:

1. Say that we know the real length of the string is 42 m.
2. There is perfect accuracy: $F = 0$ so $X = x + R$.
3. Let's say that because 42 m ÷ 0.3 m = 140 we will put the string against the ruler 140 times so we will have 140 opportunities for random error with each measurement.
4. Let's say that for each 0.3 m length there can be either an error of exactly +0.001 m or exactly −0.001 m and that the probability of each is 50%.

Now you can throw the string and the ruler away and use mathematics to model the measurement-making process. The random element can be brought in by tossing a coin or, if you have a spreadsheet program, by using a random number generator. (Creating random numbers for this exercise with a spreadsheet is covered in Example 4 in Chapter 8.) Instructions follow to take you through the generation of a single measurement from 140 random selections. You will repeat that 24 more times to acquire 25 replicate measurements (in five sets of five) and then you will calculate sample statistics as if these were real measurements.

Instructions

By following the instructions below you will fill in Table 4.3.

Table 4.3

Datum:	1	2	3	4	5	Total	
R							
X						ΣX	
$\bar{X} - X_1$							
$(\bar{X} - X_1)^2$						$\Sigma(\bar{X} - X)^2$	

$\bar{X}$ [　　　] m s^2 [　] m² s [　] m

In fact if you intend to continue studying this chapter you should copy out this table five times and repeat all the experiment five times over to produce a total of 25 data points and five sets of statistical values. If you don't have time to toss coins and you don't have a computer to generate random numbers, here are five values for R:

$R_1 = -0.010$ m

$R_2 = -0.008$ m

$R_3 = 0.014$ m

$R_4 = 0.006$ m

$R_5 = 0.010$ m

▶ Toss a coin 140 times and count the heads and tails. Heads represents an error of +0.001 m; tails represents an error of −0.001 m. Calculate your random error for your first replicate measurement by adding up the 'plus' and 'minus' errors.

$R_1 = \{0.001 \text{ m} \times (\text{no. of heads})\} - \{0.001 \text{ m} \times (\text{no. of tails})\}$

Repeat this step four times to fill the top row of the table. Each value in the top row represents the random error for each of five separate measurements. This time consuming part of the thought experiment is the part where the normal distribution has taken its effect.

▶ Fill in the second row of the table. Since we are assuming that accuracy is perfect you just need to add 42.0 m to each random error:

$X_1 = x + R_1 = 42.0 + R_1$

You now have five replicate measurements – this is called a **data sample**. It's a sample because these are just five data points and you could have made more. In real measurements you only have X values, so for the remaining steps pretend that you never knew the values of x and R. There are two statistical values you need to calculate from your replicate measurements. The mean will estimate the true length of the string and the standard deviation will estimate the precision of the measuring system.

▶ Total up the five measurements you have obtained:

$$\Sigma(X = X_1 + X_2 + X_3 + X_4 + X_5)$$

▶ Calculate the mean and put it in the box below the main table:

$$\bar{X} = \frac{\Sigma X}{5}$$

Does this mean equal the true length of the string? It doesn't have to. The mean of replicate measurements is an estimate of x only.

▶ Calculate deviations from the mean:

$$\bar{X} - X_1$$

Because you don't normally know the value of R the best you can do to estimate random variation is to accept your mean as the best estimate of x and calculate deviations of X from the mean. So fill in the third row of the table. It doesn't really matter which way round the subtraction is done because we are interested only in the magnitude of the deviation.

▶ Square the deviations:

$$(\bar{X} - X_1)^2$$

Make all the deviations from the mean positive by squaring them. This means you can look at the amount of deviation without worrying about the direction of the deviation. These values will complete the fourth row of the table.

▶ Total the squared deviations:

$$\sum (\bar{X} - X)^2$$

Add up the squared deviations for the five data points. This is a measure of variation in the data sample called the **sum of squared deviations**.

▶ Calculate the mean squared deviation from the mean. This is called the **variance** of the data and is a measure of how varied or how spread out the data sample is. Its symbol is s^2 and the squared part reminds you that the units are squared. Just divide the sum of squared deviations by 4. Statisticians argue that one should divide by degrees of freedom $(N - 1)$ rather than N. The reason for this is a little complex and beyond the scope of this text, so you have to take it on trust. (Degrees of freedom were explained in Chapter 2.)

$$s^2 = \frac{\sum (\bar{X} - X)^2}{N - 1}$$

▶ Calculate the sample standard deviation. The trouble with variance is that the units don't match the original measurements – they're squared. Find the square root of the variance. This is the standard deviation s. The standard deviation has the same units as the original measurements. In plain English the standard deviation is a measure of the average deviation from the mean. To be more correct it is the root mean square deviation from the mean.

$$s = \sqrt{s^2}$$

To summarize, you now have a sample of five measurements, each affected by random errors, and you have calculated two sample statistics from them. You have a mean, which is the best estimate the data provide of the true value of the length of the piece of string, and you have the sample standard deviation which is a measure of how far the measurements deviate from the mean. If you intend to study the later section of this chapter on the *t* distribution you should plot out another four copies of Table 4.3 and repeat nine steps four more times to obtain another four data samples.

Analysis

Figures 4.3 and 4.4 show some actual results obtained by students. You can judge whether your results are consistent with the analysis. Figure 4.3 shows how frequently measurements of each length occurred in an experiment where 200 measurements were taken (we're ignoring the means and just lumping all the data in together).

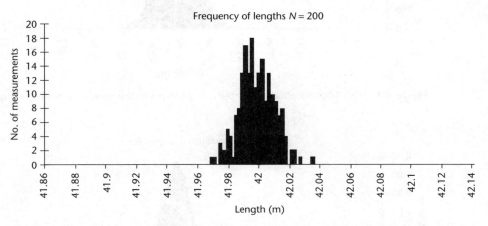

Figure 4.3 You can see that values close to 42 m are most frequently obtained and that the frequency tails off at about 42.02 m and 41.98 m.

This picture makes sense because of the way that coin throws were used to obtain each measurement. Any combination of heads and tails from 140 throws is equally possible and so the probability of each combination is 0.5^{140}. There is only one way to get 140 heads so that is an almost impossible outcome, but there are about 9.4×10^{40} different sequences of 70 heads and 70 tails that are possible, that all result in the same value for X (42 m) and this is the most likely outcome.

The probability of getting exactly 42 m is about 7%. If 10 000 measurements are made the frequency chart becomes more regular. See Figure 4.4.

Figure 4.5 shows relative frequencies that would be obtained if an infinite number of measurements were made. These are objective probabilities because they are calculated from an understanding of the system rather than from observations. Actually, you may have realized that this is in essence a binomial distribution and follows the same principles as the six-coin example presented earlier. With such a large number of coin throws all added up, the binomial distribution starts to look like a normal distribution. The probabilities of obtaining measurements of 41.998 m, 42 m and 42.002 m are almost equal. Probabilities drop with higher and lower values. Even very high and very low measurements are possible, but very unlikely.

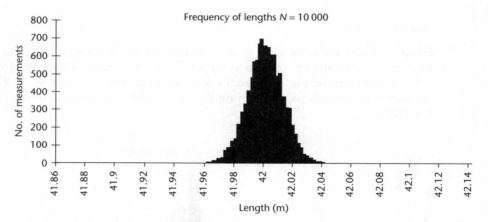

Figure 4.4 With 10 000 replicates a distinct shape forms in the frequency distribution graph.

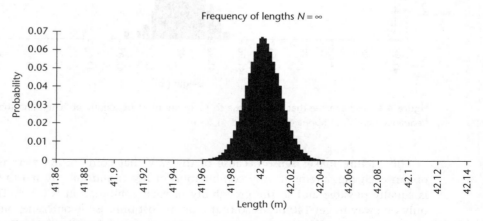

Figure 4.5 This is a graph of objective probabilities that corresponds to relative frequencies obtained from an infinite number of trials.

Population statistics and the normal distribution

The normal distribution can tell you the probabilities of obtaining certain measurements. In the string measuring example matters were simplified so that errors were always whole numbers of millimetres, but in real measurements and real error a continuous range of errors is possible. So the normal distribution has the same bell shape as the binomial distribution but the normal distribution allows a smooth range of values. There are two numbers that control the position and spread of the values. The population mean μ is the mean that you would obtain if you made an infinite number of replicate measurements $\mu = x + F$. The population standard deviation σ is the standard deviation of an infinite number of replicate measurements. The shape of the graph is the same for measurements of any type but its position is dictated by μ and its width by σ. It is possible to create a standard graph for the normal distribution that shows the spread of measurements when $\mu = 0$ and $\sigma = 1$.

Figure 4.6 tells you how likely it is that you will obtain a measurement of a certain value. You can see that the most likely measurement is 42 m but a measurement of ± 0.1 standard deviations is almost equally likely. A measurement of ± 1 standard deviation is about half as likely and a measurement of ± 3 standard deviations is very unlikely. The tails of the graph stretch off to infinity and minus infinity so theoretically it's possible to obtain any measurement although it's very unlikely that you will get measurements greater than ± 4 standard deviations from the mean.

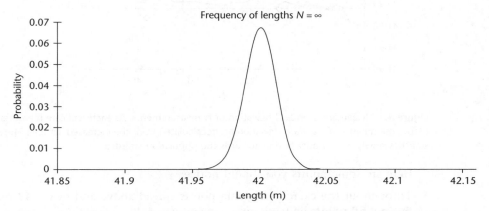

Figure 4.6 This is the probability distribution of lengths that would be obtained if measurements were governed by the normal distribution.

If you choose a range of values you can use the normal distribution to find the probability that the next measurement will fall within that range. The probability is represented by the size of the area under the graph within that chosen range. About 68% of measurements lie between +1 and −1 standard deviations; 90% of measurements lie between +1.65 and −1.65 standard deviations; 95% of measurements lie between +1.96 and −1.96 standard deviations; and 99% lie between +2.58 and −2.58 standard deviations.

Sample statistics

The mean of a sample of replicates (the sample mean) $\bar{X}$ is an estimate of the population mean μ:

$$\bar{X} = \mu + \bar{R}$$

When $N \approx \infty$ then $\bar{R} \approx 0$ and so

$$\bar{X} \approx \mu$$

The standard deviation of a sample of replicates (the sample standard deviation) s is an estimate of the population standard deviation σ. The reliability of these estimates improves as the number of replicate measurements increases.

In Figure 4.7 you see the effect of calculating a mean from increasing numbers of replicate measurements. The mean and standard deviation were recalculated each time a new measurement was made.

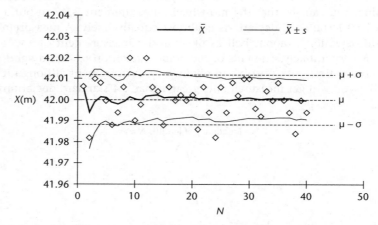

Figure 4.7 Mean and standard deviation of N measurements. As each replicate measurement is made, the mean and standard deviation are recalculated from the increased sample. The sample statistics slowly and erratically drift towards the population statistics.

Here are some points you should notice:

1. Throughout the exercise the data points spread above and below 42 m. Most but not all points lie between $\mu - \sigma$ and $\mu + \sigma$.

2. When there are only a small number of measurements the sample mean $\bar{X}$ bounces up and down above and below the population mean 42 m.

3. With more replicate measurements $\bar{X}$ settles down and drifts quite close to the population mean although it does still waver a little.

4. The sample standard deviation s becomes close to the population standard deviation σ after only a few measurements.

5. Improvements in the estimates are fast at first but then progress becomes slower and slower.

We return to these string data later in the chapter and examine the five means that you calculated in order to look more deeply at the statistics.

Lactate dehydrogenase

Now let's look at a more scientific type of replicate measurement and try to extend the thought experiment into a real-life scenario. (You might like to refer back to the lactate dehydrogenase assay protocol in Chapter 3 before continuing.) The true concentration of a lactate dehydrogenase solution and the accuracy of the assay will determine the population mean μ. The precision of the assay will determine the population standard deviation σ. In the string measuring system the key to ending up with a normal distribution was the many (140) chances for tiny errors. What is the parallel to that in the lactate dehydrogenase assay? In the assay there are several different types of random error but let's concentrate on just one – the change of temperature when a sample of the solution to be assayed is drawn into the automatic pipette.

During the fraction of a second it takes for the technician to remove his or her finger from the plunger of the pipette there are millions of little moments that make tiny changes to the temperature. If the technician grips the side of the pipette a little harder it will warm up; if the technician leans forward out of the sunshine it will cool slightly. If the temperature of the liquid differs from ambient temperature the swirling air in the pipette will be cooled by it. If the pipette was left in a cold place it will be warming up from the time it was picked up – and how long it was in the technician's hand before being lowered into the sample solution will add to the randomness. If the technician kept the pipette tip in his or her pocket . . . and so on. All of these events are like the tiny pluses and minuses in the string measuring system – and this only covers random events to do with the temperature of the pipette. To this you can add fluctuations of temperature in the reaction vessel. In most measuring situations where there are many opportunities for error, the distribution of error follows the normal distribution.

So the same statistics apply to lactate dehydrogenase assay results as were applied to the string measuring model and explained in the following pages. If 50 replicate measurements are made $\bar{X}$ will be a very close estimate of μ and s will be a very close estimate of σ. So even though you don't know the two population statistics that control the distribution of measurements, the sample statistics from a large number of measurements are practically as good. Having established that lactate dehydrogenase assay results are governed by the normal distribution, you can use the normal distribution to find the probability of your next measurement falling within any range you choose.

Confidence limits

We've now covered sufficient statistics to provide you with a way to communicate your confidence in a measurement at least in one specific situation. A knowledge of the normal distribution allows you to:

1. Make a large number of trial measurements.
2. Estimate the precision of your assay by calculating the standard deviation.

Now you can put confidence limits on single (not replicated) measurements. The normal distribution tails off at the top and bottom so it's difficult to think of an obvious range of values you can use to communicate your confidence in your estimate. You have to resort to probabilities – for any range of values you can use the normal distribution to find the probability of the true value lying within that range. There are some assumptions to make about your assay:

1. Measurements are perfectly accurate.
2. Measurements are normally distributed.
3. The precision is the same for high values and low values.
4. When you assay routinely you will only make a single measurement (no replicates).
5. Your precision will remain the same for your routine individual measurements as it was when you made replicate measurements so as to estimate standard deviation.

Just as a reminder:

$$X = x + F + R$$

When you make routine measurements you don't know x and we're assuming that $F = 0$. We don't know what R is but we have an estimate of the standard deviation s, which can be used to calculate confidence limits. You can be 68% confident that the true value lies within one standard deviation of your estimate. So, your estimate is X, your upper 68% confidence limit is $X + s$ and your lower 68% confidence limit is $X - s$. The probability of 68% is not very inspiring and 95% confidence limits are more conventional; your upper 95% confidence limit is $X + (1.96 \times s)$ and your lower 95% confidence limit is $X - (1.96 \times s)$.

This method of finding confidence limits (including the 1.96 constant) will work for any type of measurement that fits the assumptions listed above. However, there are real problems with those assumptions. The biggest problem is that you are running a special trial using many replicate measurements to try to determine the precision and hoping that the precision is maintained for the individual measurements. The next problem is that you are limiting routine assays to a single assay run when you know well that if you took the mean of several replicate measurements you would get a much better estimate. In truth, this method of putting confidence limits on a single measurement has been described here only so that you are familiar with the idea of confidence limits before we get into the t distribution.

The t distribution

The normal distribution describes the probability of obtaining a single measurement that deviates by a given amount from the true value. However, since you must make replicate measurements to estimate standard deviation, probabilities connected with single measurements are not always useful. What you really need is a probability that tells you how close the mean is to the true value. That's what the *t* distribution is for.

Thought experiment

Let's go back to the string measuring thought experiment from earlier in the chapter. What we want to do is get some kind of measure on the way that random variation in individual measurements gives rise to random variation in sample means. If you completed the whole set of calculations you should have 25 measurements in five groups and consequently five sample means. One wouldn't normally partition off measurements into groups like this but it's useful to compare the means. Without doing any maths, just take a look at the data and compare how spread out they are compared with how spread out the means are. You ought to see that the means are more consistently close to the true value than the individual measurements. (You might have been very unlucky and got data that contradict this.) This is all very well but we really need to quantify the effect we are seeing.

You quantified the spread of your individual measurements easily enough, so why not just pretend that your five means are data points and do the same with those? Here's how.

Create another twenty values of *R* in just the same way as before. (If you don't have time use the values in Table 4.4.)

Table 4.4

2nd sample	Data sample 3rd sample	4th sample	5th sample
0.004	0.016	−0.010	0.026
−0.018	−0.010	0.016	0.002
−0.004	0.008	−0.018	0.000
−0.010	0.002	−0.010	−0.008
0.000	0.006	−0.002	−0.014

▶ Make four more copies of Table 4.3.

▶ For each set of five random errors make exactly the same calculations as before. Copy the mean and standard deviation from each set of values into the appropriate rows of Table 4.5.

▶ In the row below the one for standard deviation calculate *s* divided by the square root of 5.

LIVERPOOL
JOHN MOORES UNIVERSITY
AVRIL ROBARTS LRC
TEL. 0151 231 4022

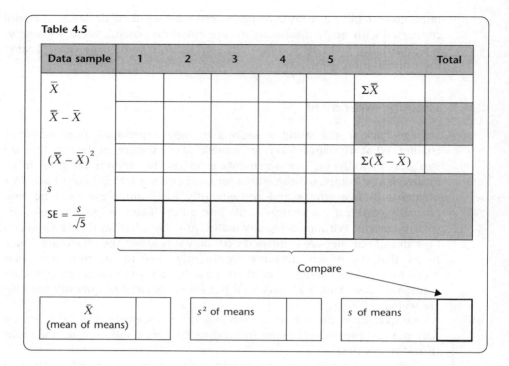

Table 4.5

Data sample	1	2	3	4	5		Total
$\bar{X}$						$\Sigma\bar{\bar{X}}$	
$\bar{\bar{X}} - \bar{X}$							
$(\bar{\bar{X}} - \bar{X})^2$						$\Sigma(\bar{\bar{X}} - \bar{X})$	
s							
$SE = \dfrac{s}{\sqrt{5}}$							

Compare

$\bar{X}$ (mean of means)		s^2 of means		s of means	

▶ For the other boxes in the table do the same calculations on means that you previously did on the individual values of X.

How far does this exercise take us? The first point to make is that this is impractical as a method of assessing the reliability of your mean. If you really made 25 measurements you would want to use all of them to calculate a single mean, not divide them up into packets. The next point is that we have no reason to assume that the means will be normally distributed, and as it happens, they aren't. But while we're on this tack, let's try another approach.

The row of figures labelled S.E. is called the **standard error of the mean** and should be similar not only among the sets of data but also to the standard deviation of means. The standard error of the mean is a very useful statistic.

Standard error of the mean

The standard error is a measure of how spread out a series of means would be if you repeated your set of replicates a number of times. The standard error would be the standard deviation of this set of means calculated as if they were data points. Fortunately, you don't actually have to repeat your set of replicates because the standard error can be estimated from the standard deviation of a single set of replicates with a single mean. This is a clever trick that was worked out by an employee of the Guinness Company, William Gosset who published anonymously as A. Student. A single set of replicates is used to reliably predict what the variation would be between the means of a number of sets of replicates. The value of the standard error is that it can be used to judge the unreliability of your mean due to imprecision.

While standard deviation is a measure of how precise an assay is, standard error is a measure of how precise a mean of a set of replicates is.

Even if you can't improve the precision of your assay you can still improve the precision of your mean by making more replicate measurements. So standard error is the statistic that should be quoted along with the mean of a set of replicates if you want to communicate how reliable the mean is as an estimate of the true value.

$$SE = \frac{s}{\sqrt{N}}$$

Shape of the *t* distribution

William Gosset worked out the distribution of sets of means either side of the true value due to random variation and called it the *t* distribution. The *t* distribution is similar in shape to the normal distribution but is distorted depending on the number of replicate measurements used to find the mean (see Figure 4.8). The other difference is that the width is governed by the standard error instead of the standard deviation. The *t* distribution governs the probability of obtaining a set of replicates with a certain mean.

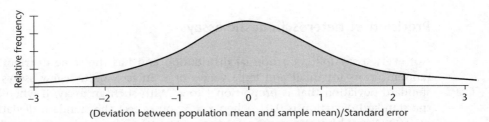

(Deviation between population mean and sample mean)/Standard error

Figure 4.8 The *t* distribution for ten degrees of freedom. The *t* distribution is similar to the normal distribution. The shaded area corresponds to 95% of the total area under the graph.

With eleven measurements the probability is 95% that the mean will lie between +2.228 and −2.228 standard error units (no matter what measuring system is involved). This corresponds to the area in the middle of the distribution that is 95% of the total area. If you don't know the true value you are measuring you can reword this as: there is a 95% chance that the true value lies within 2.228 standard error units of the mean. Statistical tables and computers usually provide computations that refer to the two tails of the distribution: there is a 5% chance [a probability of 0.05] that the true value lies outside of 2.228 standard error units from the mean. Distances from a mean in units of standard error are called *t*.

Confidence limits

You use the t distribution to find a range either side of the measured mean where the true value is likely to lie. Here's how it's done.

▶ Find the mean and the standard error of the data.

▶ Pick a probability. Most often 95% or 99% confidence are chosen which correspond to two-tailed probabilities of 0.05 and 0.01.

▶ For a particular number of measurements the t distribution tells you upper and lower limits in standard error units beyond which there is a 5% probability that the true value lies. This deviation can be called $t_{0.05}$. You can use a computer or a table of values to find these limits.

▶ Convert the t value to the units of the original measurements. Just multiply by the standard error and add to the mean to get the upper confidence limit of the mean:

$$X_{upper} = \bar{X} + (t_{0.05} \times SE)$$

▶ The lower confidence limit of the mean is calculated in a similar way:

$$X_{lower} = \bar{X} - (t_{0.05} \times SE)$$

You can say that there is a 95% probability that the true value lies between X_{lower} and X_{upper} or there is a 5% probability that the true value lies outside these limits.

Precision of heteroschedastic assays

Not every assay follows a normal distribution and has the same distribution of random errors for small and large values of x. In fact, quite a few assays show standard deviation that is proportional to x. With such an assay, if when $x = 1$, the standard deviation is 0.01, when $x = 2$ you expect the standard deviation to be 0.02. A graphical comparison of some calibration curves with 95% confidence limits added will be useful. In Figures 4.9 and 4.10 the x-axis is the concentration of the standard solution that was assayed and the y-axis is the corresponding activity of the enzyme.

When standard deviation is constant for low and high values the assay is called **homoschedastic** (Greek for same scatter), and if standard deviation varies the assay is called **heteroschedastic** (Greek for different scatter). What do you do if your assay is heteroschedastic? If you have discovered that standard deviation is proportional to amount you can express standard deviation as a percentage. If x is 1 and standard deviation is 0.01 you can express standard deviation as 1% and this can be applied to smaller and larger x values. With heteroschedastic assays the distribution of random variation is less likely to conform to a normal distribution and there may even be an asymmetric distribution. We return to heteroschedasticity in Chapter 6.

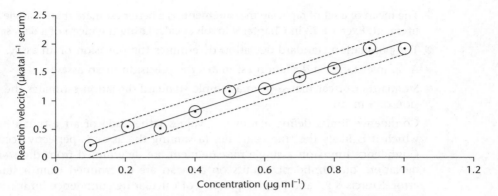

Figure 4.9 Calibration curve for assay of enzyme Y. Random variation in reaction velocity is constant for high and low values. The dotted lines indicate 95% confidence limits on measurements of reaction velocity.

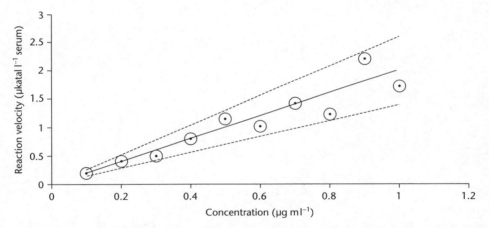

Figure 4.10 Calibration curve for assay of enzyme X. Random variation in reaction velocity is greater for higher values, as shown by the 95% confidence lines.

Summary

- The precision of an assay determines the usefulness of the assay results for decision making. Exercise 26 in Chapter 9 challenges you to explain this point with reference to a specific example.

- The random errors that cause imprecision in an assay often conform to a normal distribution.

- The normal distribution often occurs when many random events contribute to the total random error. Example 4 in Chapter 8 provides instructions for using a spreadsheet to model imprecision.

- The mean of a set of replicate measurements is a better estimate than a single measurement. Exercise 23 in Chapter 9 involves calculating the mean of a data sample.
- The population standard deviation determines the precision of an assay.
- A sample standard deviation estimates the precision of an assay.
- Standard error calculated from a sample standard deviation estimates the precision of a mean.
- Confidence limits define a range of values either side of an estimate within which it is likely the true value lies (assuming the assay is perfectly accurate). Confidence limits on a single measurement are determined from the standard deviation, but confidence limits on a mean are determined from a standard error. Exercises 22 and 24 give practice of calculating confidence limits.
- The t distribution governs the distribution of sample means.
- Some assay results are not governed by the normal distribution and don't have equal precision for high and low values. Exercise 25 demonstrates this point.

Chapter 5

Accuracy

Introduction

This chapter looks at accuracy and some examples of how to cope with it experimentally or statistically. Two significance tests will be described in the context of accuracy but both of these could be used in other situations. Inaccuracy is a fixed error, which is the same each time the measurement is repeated. As such, it is potentially a more serious problem than imprecision because its effects remain even if you make repeated measurements. On the other hand, it is often easier to eliminate inaccuracy by improving laboratory practice than it is to overcome imprecision. Put into symbols:

For one measurement, $X = x + F + R$

For a set of replicates, $\bar{X} = x + F + \bar{R}$

where x = the true value

X = the measurement

$\bar{X}$ = the mean measurement

F = the fixed error or inaccuracy

R = the random error

$\bar{R}$ = the mean random error

As the number of replicate measurements increases, $\bar{R}$ tends towards zero and so the effects of imprecision are reduced. However, F remains the same (by definition) and so when x is unknown, F is also unknown and these two values can't be disentangled. There are two basic strategies for dealing with inaccuracy: (1) control it with experimental design, and (2) detect or quantify it. We start with a short discussion of basic experimental design for controlling accuracy, then we look at the use of a standard solution in an assay validation exercise that quantifies inaccuracy and finally at the use of block design to detect inaccuracy. The two statistical methods described are the t test and analysis of variance.

Controlling accuracy

If you want to control accuracy this means that you try to design an assay so that if part of it goes a little astray it is compensated for by some other element of the experiment. For example, take the lactate dehydrogenase assay described in Chapter 3. One source for inaccuracy occurs if a series of measurements is made using degraded NADH. Every measurement will be shifted out by the same amount and so you need a control that will shift them back the other way by the same amount. One answer is to replot a calibration curve, using standard solutions of lactate dehydrogenase, at the start of a session of assay runs. That way the same calibration curve will compensate for the inaccuracy and the amount of compensation will be correct for the current batches of reagents. Compare the calibration curves in Figures 5.1 and 5.2.

The curve in Figure 5.1 is from an assay made using fresh NADH and the curve in Figure 5.2 is from an assay made using older and partly degraded NADH.

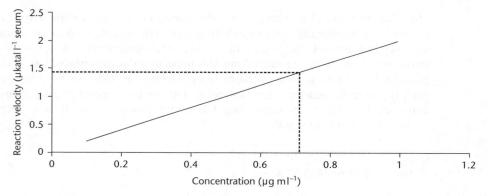

Figure 5.1   Calibration curve for an assay. The measured velocity, 1.4 µkatal l⁻¹ corresponds to a concentration of 0.7 µg ml⁻¹.

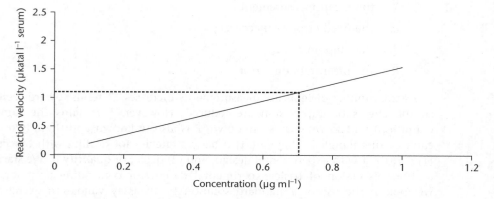

Figure 5.2   Calibration curve for an inaccurate assay. The velocity is reduced to 1.1 but the calibration curve was constructed using the same inaccurate assay, so a concentration of 0.7 µg ml⁻¹ is still obtained.

Measurements made using the degraded NADH will be too low if you leave the values as reaction velocities, but if you use the appropriate calibration curve to convert the values to the form mass per unit volume the inaccuracy will be compensated for and the correct concentration will be obtained.

There is a problem with controlling inaccuracy because the compensation may occur invisibly and you may not become aware of it when it occurs. That means that it's possible that underlying problems may become exaggerated until the point where the controls can't cope. One outcome of the control of serious inaccuracy may be that precision deteriorates. Take the example of deteriorated NADH. All reaction velocities will be much reduced and although the precision in terms of reaction velocity will be unchanged, the slope of the calibration curve will be much shallower so this will equate to poorer precision in terms of mass per volume. This point is illustrated by Figures 5.3 and 5.4.

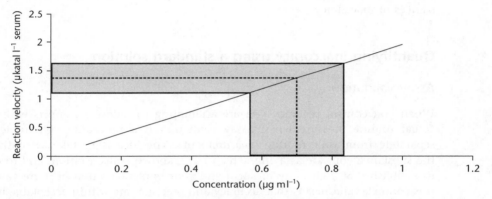

Figure 5.3 Calibration curve for an assay. The 95% confidence limits on measured reaction velocity, indicated by the horizontal band, correspond to 95% confidence limits on the concentration determined with the calibration curve, indicated by a vertical band.

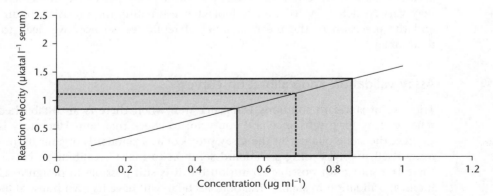

Figure 5.4 Calibration curve for an inaccurate assay. Even though the low reaction velocity will still equate to the correct mass per volume, the upper 95% confidence limit on the reaction velocity scale could equate to a higher value on the mass per volume scale, and the lower 95% confidence limit could equate to a lower value. This is indicated by the wider vertical band.

There are limits to how much inaccuracy can be controlled by small modifications to experimental design because there are sometimes many different causes for inaccuracy. In the next section of this chapter we look at how you can use a standard solution to detect and quantify accuracy. One cause for inaccuracy in the lactate dehydrogenase assay is mishandling of blood samples. It is lactate dehydrogenase in the plasma only that you wish to measure, but lactate dehydrogenase is also present in blood cells. Damage to blood cells when handling the blood sample and especially when centrifuging it to remove the cells, causes a real increase in lactate dehydrogenase in the solution you assay. The error is constant for every repeated measurement on that blood sample and so it is a type of inaccuracy. Because this inaccuracy is due to a real increase in enzyme concentration in the prepared sample it is difficult to see how we could control it or quantify it. In the last section of this chapter we will look at a type of experimental design called block design that attempts to address these more difficult sources of inaccuracy.

Quantifying inaccuracy using a standard solution

Assay validation

When you control accuracy you are attempting to reduce its effects with your actual routine measurements. Assay validation is a procedure you implement separately from your routine measurements. The idea is to take a solution of the substance your are assaying, which has a known concentration, and to use it to establish that your assay protocol and your implementation of it are valid. In this context, valid means that accuracy and precision are within acceptable limits, and to find that out, both must be quantified. The effort involved in a thorough assay validation procedure is worthwhile when your laboratory first starts implementing a particular assay protocol. You can't trust estimates of precision and accuracy quoted by other laboratories because your particular implementation may vary in detail. We've already looked at estimating the precision of an assay and the precision of the mean of a set of replicates, so now we need to look at accuracy.

Assay validation and calibration curves

This use of standard solutions will work well when there is an established and reliable assay for a substance and you want to check that your laboratory is able to meet the same quality on the assay protocol as a standards organization. If the assay itself involves using a standard solution to make a calibration curve then there is a source for possible confusion but it is still possible to design a scheme for assay validation in such a situation. Each lab will have its own house standards for frequent construction of calibration curves, but for the purposes of assay validation a centrally provided standard solution will be assayed. In a way, the process is a comparison of the house standard against a 'super' standard. There may be a difference between the way the standards are treated – there may be limited time and materials for replacing house standard solutions because they are used routinely and frequently. A 'super' standard will probably need to be made with more care. These issues about standard solutions are not important for the remainder of this section because we are going to concentrate on what you do with the measurements, regardless of how they were obtained. So, where a standard is mentioned, that would refer to the 'super' standard in an assay that uses a house standard.

Standard solutions

Our assumption is that the concentration of the standard is known with perfect accuracy and precision. If the assay to be validated had perfect precision then you would only need to make one measurement and the difference between the known concentration and the measurement would be the fixed error that determines accuracy:

$$\text{When } R = 0, \quad X = x + F$$
$$\text{i.e.} \quad F = X - x$$

However, in real world assays there is some imprecision that makes this simple procedure ineffective and so you must make replicate measurements and apply some statistics. This will involve using the same t distribution that was used in the previous chapter to calculate 95% confidence limits on a mean.

Assay validation with a t test

Here is a procedure for validating an assay using a standard solution. This is not the only way to validate an assay but it's a simple way to apply a statistical test. One aim is to decide if there is significant inaccuracy and another is to measure the amount of inaccuracy if there is any. Let's start by imagining an assay that is perfectly precise and then factor in the extra statistics needed to work with accuracy if the assay is also imprecise. Detecting and quantifying accuracy for a perfectly precise assay is easy:

$$\bar{X} = x + F + \bar{R}$$

The value of $\bar{R}$ will be zero for a perfectly precise assay, so the simplified equation is:

$$\bar{X} = x + F$$

Which can be rearranged as:

$$F = \bar{X} - x$$

Because a standard solution was used, x is known. Put into plain English, your accuracy is the difference between your estimate and the known value of your standard. A perfectly precise assay is also perfectly accurate if the estimate is the same as the known value. In other words, if the assay is perfectly precise but

$(\bar{X} - x) \neq 0$ then it must be inaccurate. Assays are not perfectly precise so, sadly, things are not as simple as this.

The consequences of imprecision

Assays are always imprecise to some degree so the maths is not as simple as you might like it to be. You can still find the distance of your estimate from the standard, but since $\bar{X} = x + F + \bar{R}$ and x is known, it follows that:

$$\bar{X} - x = F + \bar{R}$$

We have no way to break apart F and $\bar{R}$ because we have no way to exactly quantify either one separately. What is the consequence of that? With an imprecise assay, if the estimate differs from the standard you can't tell for certain why it differs. It could differ because of inaccuracy, because of imprecision or both. Put another way, if $(\bar{X} - x) \neq 0$, then F could still be zero and the assay could be accurate. This uncertainty means that we need to do a significance test. (You are advised to study the Chapter 2 if you haven't done so already.) The first requirement of a significance test is a null hypothesis, and in our case that can be expressed in mathematical symbols as:

$$F = 0$$

Put into English, our null hypothesis is that the estimate deviates from the standard *only* because of random variation in replicate measurements, i.e. because of imprecision only. The alternative hypothesis is that the estimate deviates from the standard owing to inaccuracy or a combination of inaccuracy and imprecision. This alternative hypothesis is supported if the estimate deviates significantly from the standard and the null hypothesis must be rejected. Another look at the t distribution will aid an understanding of the statistical method described below. In Figures 5.5 to 5.9 the bold vertical line represents x, which is known. The bold curve represents a t distribution around x. If the assay is accurate, this t distribution will govern the probability of getting a mean that differs from x. It is the distribution that fits with the null hypothesis. The values of replicate measurements are represented by circled points on a horizontal axis. The fine vertical line represents the mean of the replicate measurements. In some of the figures an inaccurate assay is represented and the dotted vertical line represents $x + F$ which is an unknown value. The dotted curve is the t distribution that really operated because the assay was inaccurate. It is the t distribution that *really* governs the probability of getting a mean that differs from x for the *in*accurate assays. If you want to look at these graphs from the point of view of someone actually doing the experiments, imagine that the dotted lines are not included on the graphs.

Moderately imprecise, perfectly accurate (Figure 5.5)   The important point to note here is that the mean value is higher than x. That is quite possibly owing only to imprecision because of the random nature of imprecision. We would expect that if the experiment were repeated we would get a different mean which might be lower than x.

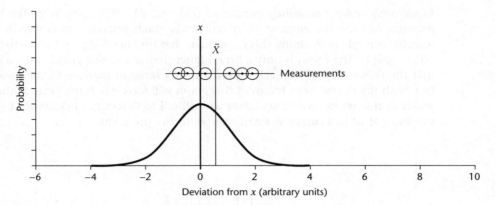

Figure 5.5   Moderately imprecise, perfectly accurate.

Moderately imprecise, moderately inaccurate (Figure 5.6)   When you actually carry out this procedure you don't actually know if the results are inaccurate, but this graph shows the 'unknown' value of $x + F$ as a dotted vertical line. This means that the real t distribution that is governing the distribution of means is the dotted one. Again you won't know that when you do a real experiment; all you will be aware of is the measurements themselves and your mean. In this case the mean is a long way above x and the individual measurements lie higher than the tail of the bold t distribution. The mean disagrees with the null hypothesis because getting this mean from the t distribution suggested by the null hypothesis would be very unlikely. Since the null hypothesis is disproved we have to accept that the assay is inaccurate. The distance of the mean from x is an estimate of the amount of inaccuracy. As it happens, the mean is somewhat higher than $x + F$ so your estimate of the inaccuracy will be a little too high, again owing to the imprecision.

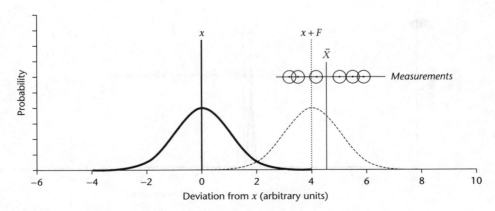

Figure 5.6   Moderately imprecise, moderately inaccurate.

Moderately imprecise, slightly inaccurate (Figure 5.7)   This figure is similar to the previous one but the amount of inaccuracy is much smaller. Remember that the scientist actually performing this experiment has no knowledge of the dotted lines on the graph. The mean is only a little higher than *x* and one could easily assume that the reason is that imprecision has led to a random increase in the mean. In fact, with the dotted lines removed this graph will look much the same as the first graph in this series. With these data we will fail to detect the inaccuracy because the amount of inaccuracy is small relative to the precision.

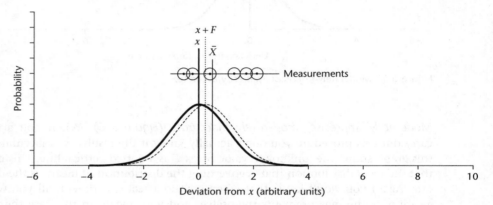

Figure 5.7   Moderately imprecise, slightly inaccurate.

Slightly imprecise, slightly inaccurate (Figure 5.8)   This figure is similar to the previous one, and the amount of inaccuracy is the same, but the assay is more precise. So the *t* distributions are narrower and the data points are closer together. This means that the data points now lie on the extreme tail of the bold *t* distribution and we can easily reject the null hypothesis. There is no *more* inaccuracy in this experiment than the previous one but it has become *more significant* because the assay is more precise and we are better able to estimate inaccuracy.

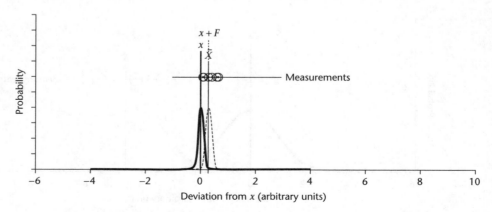

Figure 5.8   Slightly imprecise, slightly inaccurate.

Very imprecise, moderately inaccurate (Figure 5.9)   This example shows the reverse of the previous one. We now show an assay that is quite inaccurate but because it is also very imprecise this inaccuracy can't be detected. This is like Figure 5.7 but with the range of values scaled up; it's like an enlargement of the same graph. With such an imprecise assay it is quite feasible to obtain a very high mean due only to random variation caused by imprecision.

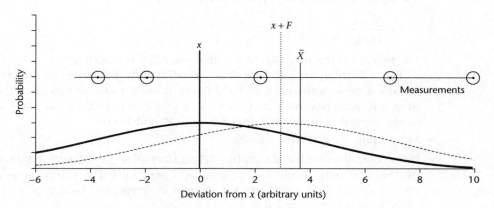

Figure 5.9   Very imprecise, moderately inaccurate.

These examples are fairly easy to interpret because the values have been chosen deliberately so that they're easy to interpret. With real assays there may be border-line cases and with many experiments performed it is inevitable that some improbable random errors will occur sometimes. For these reasons it is important that a systematic method of analysis is applied.

Step by step method

▶ Decide in advance how high the level of significance must be before you accept that inaccuracy has been detected (the critical probability). A common convention is 95% significance. If the probability that inaccuracy contributed to the deviation is greater than 95% then you assume that there is inaccuracy in the assay and take appropriate action. (See Chapter 2 for a discussion of critical probability.)

▶ Make a number of replicate measurements. More measurements would be better but there is a limit on the time and materials you would want to commit to this.

▶ Calculate the mean of the replicates.

▶ Calculate the standard error of the replicates.

Since $\bar{X} = x + F + \bar{R}$ and x is known it follows that:

$$\bar{X} - x = F + \bar{R}$$

▶ Consequently you can calculate the deviation of the mean from the known concentration of the standard $(\bar{X} - x)$. For an accurate assay this is equal to $\bar{R}$, but for an inaccurate assay it could be a much bigger (positive or negative) number.

▶ Standardize the units of the deviation – divide by the standard error. In effect this tells you how far the mean measurement is from the true value relative to the precision of the assay. This statistic is called t:

$$t = \frac{\bar{X} - x}{s/\sqrt{N}}$$

The bigger t is the more likely it is that the assay is inaccurate.

The t distribution tells you that $\bar{R}$ could easily be within one or two SE (standard error) units of 0; and if $(\bar{X} - x)$ is within one or two SE units of 0 then it is quite possible that $(\bar{X} - x) = \bar{R}$ and $F = 0$. If $(\bar{X} - x)$ is big (positive or negative) it is more likely that $(\bar{X} - x) \neq \bar{R}$ and $F \neq 0$.

▶ Make t positive if it isn't already.

▶ Use a computer or table of statistics to find the probability of getting a deviation this big due only to random variation (imprecision). The computer must be provided with the t value and the degrees of freedom (which is $N - 1$). This probability is usually called p.

Using an Excel spreadsheet, if cell A1 contained the value of t and cell B1 contained the value of N you could calculate p by entering the following formula into a cell:

=TDIST(A1, B1-1, 2)

The digit 2 in the function indicates to Excel that you want to account for both tails of the t distribution (as indicated in Figure 5.10). You need both tails because you are accepting that inaccuracy could be positive or negative. The bigger t is, the smaller p will be. In other words, the further your mean is from the true value, the less likely it is that the assay is accurate.

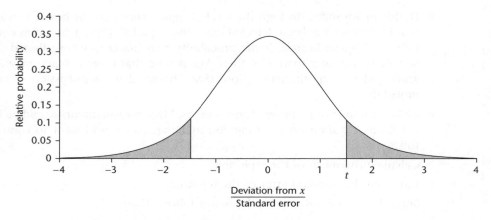

Figure 5.10 The t distribution for ten degrees of freedom.

▶ You can also calculate $1 - p$ if you want. This is the probability that inaccuracy *did* contribute to the deviation of the mean from the true value (represented by the middle part of the t distribution shown in Figure 5.11).

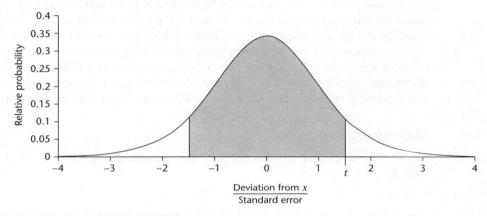

Figure 5.11   The t distribution for ten degrees of freedom.

This statistical procedure is called a t test. It tests the significance of a deviation against the random variation in the data. Statistical tests in general test the significance of statistical values and arrive at one or more probabilities. In a previous exercise you used the t distribution to put confidence limits on a measured mean. You chose a probability and found a deviation. For this significance test you do the reverse: you have a deviation and you want to calculate a probability. If you aren't much interested in the exact value of the probability you could just calculate 95% confidence limits in the usual way and if x lies outside those limits you know that $p < 0.05$ and the assay is significantly inaccurate.

🕮 *Warning, warning, warning*

There is an extremely important point that you must keep in mind when you perform this type of assay validation. The more imprecision you have in the assay the less likely it is that you will be able to detect inaccuracy. So if you wanted to deceive yourself or others that your assay was accurate, you could deliberately create as much imprecision as possible. All the inaccuracy will magically disappear along with your credibility as a scientist. (This is a bit like when Lord Nelson put the telescope to his blind eye and said 'I see no ships'.) Take another look at the five example situations in Figures 5.5 to 5.9 if this doesn't make sense.

Block design

At the start of this chapter there was a discussion of the problem caused by detecting certain sources of inaccuracy. For example, a lactate dehydrogenase assay may seem to be perfect by the assay validation procedure described in the previous section, but when you assay real blood samples rough handling could raise concentrations of lactate dehydrogenase and cause undetected inaccuracy. Block design provides a way to detect inaccuracy with actual routine measurements. In the worked example below, any inaccuracy is assumed to be due to broken blood cells in the blood samples used. The essence of block design is to make blocks of replicate measurements and in each block to repeat key parts of the preparation. So for our example the blood sample will be divided into three portions and each portion will be prepared for the assay separately, perhaps by different technicians using different equipment. The critical part of the preparation is centrifugation and you might use different centrifuges. (Low-speed centrifuges are more likely to cause swirling inside the tubes and consequent cell damage than high speed centrifuges.)

Mathematics

The mathematics that governs the replicate measurements within a block is the same as previously discussed:

$$\bar{X} = x + F + \bar{R}$$

The difference is that each block of measurements is a distinct set of replicates. Within a block the fixed error must be constant (by definition) but different blocks may have different fixed errors. This is obvious if you think about the lactate dehydrogenase example: different levels of rough handling of the portions of blood sample may result in different amounts of enzyme in the different portions of serum. Suppose there are three blocks in the experimental design, say:

$$\bar{X}_1 = x + F_1 + \bar{R}_1$$
$$\bar{X}_2 = x + F_2 + \bar{R}_2$$
$$\bar{X}_3 = x + F_3 + \bar{R}_3$$

Let's take the different elements of these equations and discuss how they may vary from block to block.

The x value

This represents the true concentration of lactate dehydrogenase in the serum of the blood before you separated out the blood cells and possibly broke some open. It must be the same for all three blocks because every block was based on a single sample of blood that was simply divided into three portions. Although we do know it is the same for all three blocks we can't actually know what value it is. (We only know the value of x in the assay validation previously described because we were using a standard solution; this time it's a real experimental blood sample.)

Fixed error F_1, F_2 and F_3

There are different possibilities for the fixed error:

1. If no damage is done to blood cells when the blood is centrifuged and no other problems with the assay then all three of these values will be zero.

2. If blood cells are damaged when the blood sample is handled the values will be non-zero and will be different from each other. They will differ because each portion of the blood was handled separately. It's possible that one or two of the fixed errors will be zero and a single block will produce a non-zero fixed error. This is the type of situation that block design can detect.

3. There may be a source of inaccuracy other than rough handling of the blood sample, for example the use of degraded reagents that will equally affect all the blocks. In this case all three F values will be non-zero but equal to each other. This type of inaccuracy cannot be detected in a block design.

Despite these conclusions about fixed error, we will never actually know the values of F_1, F_2 or F_3 because we can only calculate the block means and we don't know other terms in the equation that would allow appropriate arithmetic.

Mean random error $\bar{R}_1$, $\bar{R}_2$ and $\bar{R}_3$

The distribution of random variation will be the same in all three blocks because the assay protocol is applied in exactly the same way for all three blocks. For all three blocks the mean random variation will tend towards zero as more replicates are made, and the spread of replicates will be determined by the same underlying population standard deviation.

However, since a finite number of measurements are made in each block and since imprecision is random these three values will be non-zero and will differ from each other:

$$\bar{X}_1 = (x + \bar{R}_1 + F_1)$$
$$\bar{X}_2 = (x + \bar{R}_2 + F_2)$$
$$\bar{X}_3 = (x + \bar{R}_3 + F_3)$$

The three terms that contribute to the mean measurement in each block cannot be separated from each other, but that doesn't prevent comparisons between blocks. Imagine for a moment that we know exactly what is going on in a particular assay and we can predict the outcome. In our example the assay will be very precise so all three mean random errors will be near zero, the fixed errors for blocks 1 and 2 will be zero (i.e. perfectly accurate) but measurements in block 3 will be very inaccurate. We can construct a graph, using the t distribution to predict a possible outcome of the experiment (see Figure 5.12). The solid curve shows the t distribution that is governing means in the two accurate blocks. The centre of the t distribution sits exactly on the true concentration. Despite the perfect accuracy the two block means from the accurate blocks will be different owing to random variation. They shouldn't be very far from x because the precision is quite good. The dashed curve is the t distribution that governs the mean in the inaccurate block. The precision is the same for the third block as for the first and second, so the distribution is the same width but the inaccuracy has caused the distribution to shift up on the axis. This results in a mean measurement for block 3 that is much higher than the other two blocks' means.

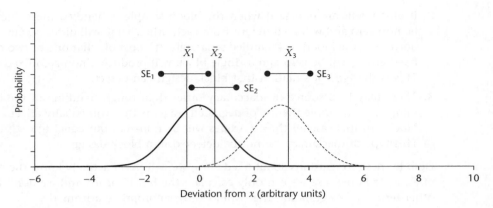

Figure 5.12   Accuracy and precision in block design.

This explanation is quite artificial because many of the elements on the diagram would not be known to the experimenter in the real situation. Figure 5.13 has all the unknown elements removed.

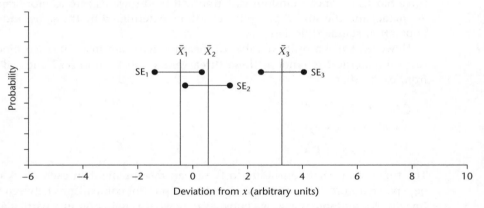

Figure 5.13   Accuracy and precision in block design. Without the benefit of knowing the true concentration you cannot be certain which blocks are inaccurate and which are accurate.

What can you tell from the data and the block means alone? With a subjective view of this example you must admit that there is some inaccuracy somewhere in the assay because there is such a wide gap between one block of measurements and the others. The most obvious explanation is that block 3 results are inaccurate and blocks 1 and 2 results are accurate, but it would be dangerous not to consider other possibilities:

1. All three blocks are accurate. This seems very unlikely from a subjective look at the data because it would require an unusual mean random error in one or more blocks.

2. All three blocks are inaccurate by exactly the same amount. This also seems unlikely from a subjective look at the data.

3. Blocks 1 and 2 are accurate but block 3 is inaccurate. This would seem to fit with the data.

4. Blocks 1 and 2 are both inaccurate but block 3 is accurate. This is feasible if blocks 1 and 2 were both done in the same, faulty way leading to a negative fixed error. (For example, both portions of the blood serum were stored in a warm place and an enzyme was degraded.)

5. All three blocks are inaccurate but by different amounts. For example, bad storage may have degraded lacate dehydrogenase in all the blocks, giving a negative fixed error but the portion of blood used for block 3 was also handled roughly which caused a positive fixed error.

A subjective look at the data would seem to reject hypotheses 1 and 2 but hypotheses 3, 4 and 5 all fit the data. You could produce a hypothesis that is a more general description of hypotheses 1 and 2:

The three blocks have equal fixed error.

You could do the same for hypotheses 3, 4 and 5:

The three blocks have differing fixed errors.

We have a situation where there is a null hypothesis, and it is random variation that makes it difficult to make an immediate conclusion. You should recognize that a significance test is needed at this point. This will start with the raw data and lead to a probability, which you can attach to the null hypothesis. In symbols the null hypothesis is:

$$F_1 = F_2 = F_3$$

Which implies that variation between $\bar{X}_1$, $\bar{X}_2$, $\bar{X}_3$ is due only to variation between $\bar{R}_1$, $\bar{R}_2$, $\bar{R}_3$.

Implementing block design

The first step to block design is to identify critical points in the assay that might result in inaccuracy. So far we've identified handling of the blood samples as a possible problem, but there may be other problems. (If you're using degraded reagents the resulting inaccuracy might be compensated for by the use of a calibration curve in the assay protocol so it may not be necessary to factor that into the block design. On the other hand, if it is considered too expensive to redo calibration curves with every new batch of reagents you might use different batches of reagents with each block but use the same lab calibration curve for all the blocks.) Once you have isolated possible problem areas, then:

1. A set of replicate measurements is made.

2. All of the parts of the assay that might contribute to uncontrolled inaccuracy are duplicated. So, for example, blood samples are divided into portions and processed separately, fresh reagents are made up, or alternative equipment is used.

3. Another block of replicate measurements are made.

4. A third or even a fourth block of replicate measurements may be made using fresh reagents etc.

5. A significance test is applied to the data.

6. If there is significant variation between block means then all the data are rejected as inaccurate.

7. If there is not a significant variation between block means then all the data are pooled and a single mean calculated which is your final estimate of the concentration.

It would be dangerous to be complacent when the null hypothesis is confirmed, because the unhappy situation of equally high inaccuracy in all blocks is one possibility allowed within the null hypothesis. If the means of all blocks agree with each other then there can be equal non-zero inaccuracy in all blocks and this cannot be distinguished from zero inaccuracy in all blocks.

Analysis of variance (AnoVa)

A statistical test called analysis of variance (abbreviated to AnoVa) can calculate from the raw data the probability p that block means vary owing to random variations only. Your computer can take the data and give you a probability without you having to understand the statistics at all, but a worked example follows to show you what the computer is doing. The basic principle is the same as any other significance test:

1. Decide on a null hypothesis.

2. Calculate various figures from the raw data using a method appropriate to the null hypothesis.

3. Calculate the probability of the null hypothesis.

4. Take action depending on whether p is above or below a predetermined critical probability.

With AnoVa you can get a spreadsheet program such as Excel or a specialist statistical package to do all of the calculations in one go and you can just read off the probability.

Worked block design example

Four sets of five replicate measurements were made of serum lactate dehydrogenase using four separate sets of reagents (shown in Table 5.1). Units are μmol s^{-1} l^{-1} but the spreadsheet takes no notice of units and they will be omitted from the remaining text.

Table 5.1

A	B	C	D
3.07	2.99	2.90	2.93
2.97	2.89	3.22	3.16
2.94	3.00	2.80	2.89
3.07	2.97	2.99	3.10
2.98	2.85	2.93	3.09

The Data Analysis command of Excel was used to perform a single factor AnoVa and produced the following two tables of figures. The first, Table 5.2, gives summary statistics for each of the four groups of replicates; the second, Table 5.3, contains the analysis itself.

Table 5.2

	A	B	C	D	E
1	Groups	Count	Sum	Average	Variance
2	Column 1	5	15.03614	3.007227	0.003642
3	Column 2	5	14.68633	2.937267	0.004269
4	Column 3	5	14.82527	2.965054	0.024293
5	Column 4	5	15.16992	3.033983	0.014137

Table 5.3

	A	B	C	D	E	F	G
1	Source of variation	SS	df	MS	F	P-value	F crit
2	Between groups	0.027833	3	0.009278	0.800836	**0.511422**	3.238867
3	Within groups	0.185359	16	0.011585			
4							
5	Total	0.213192	19				

Only the figure in Table 5.3 labelled *P*-value, which contains the probability that the four block means are due only to random variation (imprecision), is important for coming up with a final conclusion. All the other figures are just intermediate statistics that lead to the probability value. Single-factor AnoVa partitions variation into different sources. For each group of replicates there is the usual random variation, owing to imprecision within the groups. There is also variation between the means of the groups of replicates. Ideally the variation between groups should be about the same as the variation within groups because both will be due to imprecision only. However, if the different blocks have different accuracy the variation between group means will be higher.

Between groups

This deals with variation between group means. The first step is the sums of squares of deviations of group means from the overall mean (SS = 0.027833 in Table 5.3). This treats the group means as if they were four data points and calculates sums of squares in the ordinary way. The table goes on to show mean squares (MS = 0.009278). In effect this is another word for variance since it is just the sums of squares divided by the degrees of freedom (d.f. = 3).

Within groups

This deals with variation within groups. The sums of squares values is simply the total for each of the individual block sums of squared deviations (SS = 0.185359). Again the mean squares (MS = 0.011585) is like a variance because it is just the sums of squares divided by degrees of freedom (d.f. = 16).

Total

The total sums of squares could be calculated two ways. You could take all the measurements from all groups, treat them as one big set of replicates and do an ordinary sums of squares calculation. The simpler method is just to add the two contributing sums of squares values that you already have. Do a quick check on the figures in Table 5.3: do sums of squares for between groups and sums of squares for within groups add up to give the correct total?

The test

To make the test the two variances must be compared. This is done simply by calculating the mean squares ratio of 'between' to 'within'. This is called the f statistic. Because both mean squares are in the same units and f is a ratio, comparison and standardization are done in one step. If f is big then variation between groups is high compared with within groups. An f value of roughly 1.0 is expected if there is no significant difference between group means. From the table $f = 0.800836$.

The f distribution describes the probability of obtaining an f value of a certain size by imprecision alone. So the next value in the table is the p value. This is the probability that the f value could have occurred because of random variation in the replicates. If p is below a chosen threshold (often 0.05), then you accept that there is different accuracy occurring between different sets of measurements. In this test $p = 0.511422$ which is much greater than 0.05 so there is not a significant difference between groups. It's worth pointing out that as f increases p decreases.

The last value in the table is labelled f_{crit}. This is the critical f value. It is the f value that would give a p value of 0.05. Because there is not a significant difference between the block means we can treat all the measurements as a single set of replicates and calculate a mean.

Summary

- Inaccuracy is a systematic error in assays.
- Controlling accuracy means modifying the assay protocol so that it incorporates compensations for any inaccuracies.
- Experimental methods can be devised for detecting or quantifying inaccuracy.
- Assay validation is the process of establishing the expected precision and accuracy of a new assay protocol or a new implementation of an existing assay protocol.
- Imprecision in an assay makes the detection and quantification of accuracy more difficult.
- One way to quantify inaccuracy is to assay a standard solution. In this type of experiment a significance test, the *t* test, can be used to determine the significance of the inaccuracy. Example 6 in Chapter 8 demonstrates how to put together a spreadsheet for implementing this significance test. Exercise 27 in Chapter 9 tests your ability to interpret the results of such a test. Exercise 28 asks you to do the calculations of a *t* test, except the determination of a probability, with a pocket calculator, and Exercise 29 ask you to do the last part of the process with a computer and complete a statement of the result.
- Block design is a type of experimental design that can be used to detect inaccuracy in routine experiments, i.e. without the aid of a standard solution. In this type of experiment a significance test, an AnoVa, can be used to determine the significance of the variation between block means. Example 8 in Chapter 8 shows how to analyse data from a block design experiment. Exercise 30 in Chapter 9 asks you to interpret the results from this type of analysis and Exercise 31 asks you to perform the analysis with a spreadsheet.

Chapter 6

Relationships between variables

Introduction

A large part of science involves looking at the relationships between the different properties of systems. Here are some examples:

1. A raised concentration of lactate dehydrogenase in the blood plasma is an indicator of cell damage somewhere in the body because the enzyme is normally found only inside the cell. As it happens, there are different isoenzymes of lactate dehydrogenase in different tissue types. (Isoenzymes are proteins with different structures that catalyse the same reaction. They may differ only slightly in their primary structure.) These isoenzymes have slightly different reaction kinetics so there is a good reason to want to research the relationship between substrate concentration and initial reaction velocity. This research might lead to insights on why different isoenzymes have evolved for different tissue types. It might also be possible to design assays that can distinguish between tissue damage in different organs of the body. Part of your research might study the relationship between the concentration of one of the substrates – pyruvate – and the initial reaction velocity.

2. When a drug is injected intravenously it may start to take effect immediately, but how long will it continue to work? Drugs are often metabolized and become inactivated or they might be excreted, so you would be interested to know the relationship between time and drug concentration in the blood.

3. When you invent a new drug you have to decide on appropriate doses. To do that you need to find some way to measure the effect of the drug and then find the relationship between the dose and the subject's response to the drug.

4. Imagine you have found a new strain of *E. coli.* You might be interested to know about its ability to grow. That would involve incubating a sample in suitable conditions and measuring the growth of the culture. Here you are looking at the relationship between dry mass of the culture and time.

5. There are some physical aspects of biological systems that are also of interest. For example, the relationship between the internal surface area of a human lung and its length.

In each of these cases you would devise an experiment where you measure two values for each data point. The properties of a system that change in your experiment are called variables. In some cases (examples 1–4 above) you are able to fix one variable to a value you choose and then measure the other variable. For other experiments (example 5 above) you can't set either variable and you have to measure both variables for a number of subjects. The results of these experiments are series of data points where each datum has two numbers, usually referred to as x and y. When you perform this kind of experiment you need to find the relationship between x and y either graphically or statistically. In the rest of this chapter the phrase 'line fitting' will be used to refer to both the graphical and statistical methods.

There are two objectives of line fitting:

1. *To determine the mathematics that defines the relationship (if there is a relationship).* For example, in the case of the enzyme kinetics example you can find out two constants, called V_{max} and K_m, which will be different for different isoenzymes. Looking at the mathematics of the relationship could be used to try to identify the isoenzyme.

2. *To provide a way to predict the value of one property from the value of the other.* When you create a calibration curve for an assay you relate the output of the assay to a concentration in, for example, mass per unit volume. You are less interested in the mathematics that describes the relationship – you just want to use the line you fitted as a laboratory tool.

In this chapter the principles of least squares line fitting are described. This includes linear, non-linear, weighted and non-weighted methods but concentrates on the validity of methods applied to different types of data and the interpretation of results rather than the mathematical methods. Line fitting is usually applied using computer software, and knowledge of the principles is more important than memorizing the mathematical operations involved. Having said that, the basis of all least squares line fitting methods is the residual, and so residuals are looked at mathematically as well as in terms of their practical use in line fitting.

Types of relationship

The methods you use to study a relationship will depend on the type of relationship. The following are a summary of the technical jargon used to describe relationships between variables. The issues raised briefly here will be dealt with in more detail as the chapter progresses.

Linear versus non-linear

The relationship between two variables is linear if the graph of the relationship is a straight line and the mathematical description of the relationship is of the form:

$$y = mx + c$$

If you are rusty on algebra you may not recognize a linear relationship from some equations. It's a linear relationship so long as the constants in the equation can be rearranged to something like the above. The constants m and c could each be expressions involving several constants. For example:

$$y = \frac{x + c}{m + \pi}$$

can be rearranged to:

$$y = \frac{1}{m + \pi} x + \frac{c}{m + \pi}$$

which more obviously corresponds to the familiar equation of a straight line.

If the relationship between x and y can't be described by the equation of a straight line then the relationship is non-linear. For example:

$$y = x^2 + c \quad \text{or} \quad y = \frac{mx}{x - 1}$$

Bivariate normal versus univariate normal

There are some experiments where the value of x is chosen by the experimenter and is determined with close to perfect accuracy and precision. This is true of the various experiments that plot a change over a period of time. The value of y, however, is often subject to the normal distribution because it is the result of a relatively imprecise assay. In this type of experiment x is called the **independent variable**, y is called the **dependent variable** (because its value depends on x) and the data are **univariate normal**. This technical term means that one variable is subject to variation determined by the normal distribution. In other experimental situations both the x and y values are normally distributed either because both measurements are subject to imprecision or because you are looking at different individuals affected by random influences. In such cases the data are called **bivariate normal** and the variables are not labelled as dependent or independent. The choice and validity of statistical methods depends on which of these categories the data fit into.

Homoschedastic versus heteroschedastic

If the precision of the assay of a variable is the same for high and for low values it is called **homoschedastic**. If the precision of the assay of a variable varies for high and for low values it's called **heteroschedastic**. Some types of homoschedastic data are quite common for certain assays. For example, random variation may increase in proportion to the magnitude of the value being measured. Compare plots of y against x in Figures 6.1 and 6.2.

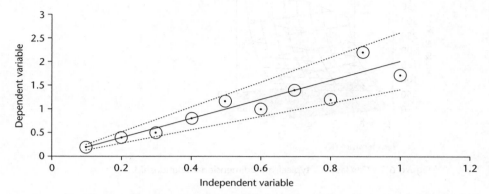

Figure 6.1 Independent variable is heteroschedastic. The y values are more scattered on the right-hand side than the left. Scatter is shown with divergent dotted lines.

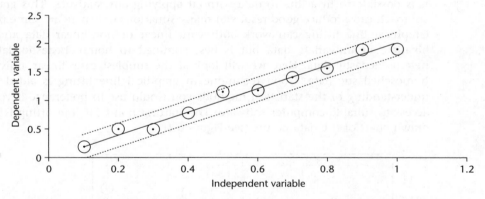

Figure 6.2 Dependent variable is homoschedastic. The y values are scattered equally on left and right. Scatter is shown with parallel dotted lines.

The choice and validity of statistical methods depend on whether the data are homoschedastic or heteroschedastic.

Multidimensional versus two-dimensional

In many experiments you determine the relationship between two variables that can be represented on a two-dimensional graph by a line. There are occasions where three or more variables are studied and the data from such a study are called multidimensional. To represent a relationship between three variables graphically you can use a surface within a 3D perspective graph or use contour lines on a 2D chart (see Figure 6.3). There are statistical methods for studying multidimensional relationships but they go beyond the scope of this text.

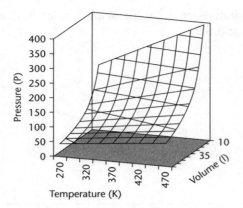

Figure 6.3   Gas law – a typical multidimensional relationship.

Graphical line fitting – linear data

It is possible to fit a line to data without applying any statistics. This graphical approach can produce good results in many situations but there are some pitfalls. Graphical line fitting can work both with linear or non-linear data and with bivariate or univariate data but is best confined to homoschedastic data not heteroschedastic data. Here we will look at the simplest case: linear, univariate, homoschedastic data. Good technique in graphical line fitting is aided by an understanding of the statistical methods you would use in preference if you had access to suitable computer software. The starting point for line fitting is a well drawn chart of the data points (see Figure 6.4).

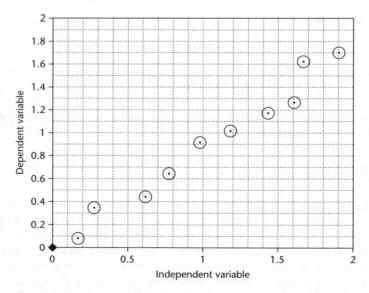

Figure 6.4   It is convenient to indicate points with dots and to remind yourself of imprecision by surrounding them with a circle.

It is at this point that you must think about the nature of the relationship between x and y. Can you assume that there is a linear relationship between them? If you can, that will simplify the line fitting because you can use a ruler. Alternatively, you can make use of a flexicurve. The basic aim of line fitting is to minimize the distances of the data points from the line without making a curve wiggle up and down to achieve that. Figures 6.5 to 6.8 show some examples of well fitting and ill fitting curves and straight lines.

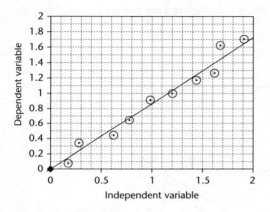

Figure 6.5 Well fitting line.

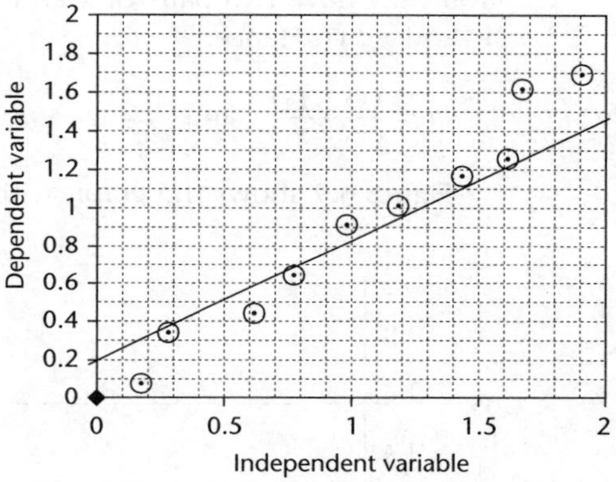

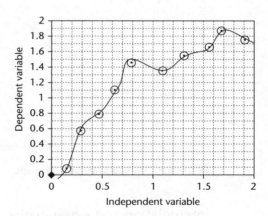

Figure 6.6 Ill fitting line.

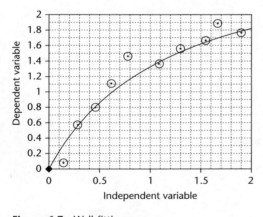

Figure 6.7 Well fitting curve.

Figure 6.8 Ill fitting curve.

Defining the relationship

If you can assume a linear relationship between x and y you need to determine m and c in the following equation to fully define that relationship:

$y = mx + c$

You can use the line you fitted to the data by eye to estimate m and c. Here's how:

▶ Pick two values for x at opposite ends of your graph. It is convenient to use values that correspond to vertical lines on your graph paper. (If your graph allows, x_1 might be zero, in which case the maths is simplified a little.)

▶ Run vertical lines from these two points on the x-axis up to the line you put through the data.

▶ At the points where the vertical lines hits your fitted line run horizontal lines over to the y-axis.

You now have two pairs of x and y values that define two points on the line, x_1, y_1 and x_2, y_2. Hence:

$$m = \frac{(y_2 - y_1)}{(x_2 - x_1)} \quad \text{and} \quad c = y_2 - mx_2$$

Figure 6.9 shows the graph with construction lines added.

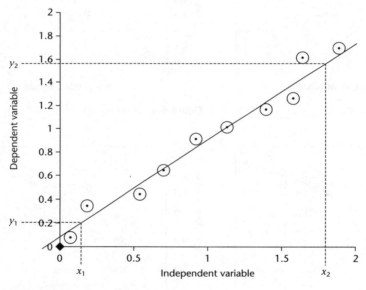

Figure 6.9   Finding slope and intercept.

Interpolation and extrapolation

From Latin, interpolation means painting between, and extrapolation means painting beyond. Graphically, interpolation means drawing the line between the data points and extrapolation means extending that line beyond the highest and lowest x values in your experiment. You can also interpolate and extrapolate mathematically using the values of m and c you calculated. Simply plug your chosen value of x into the equation of a straight line along with the m and c constants to find y. Interpolation is not a great problem but you should be very wary of extrapolation. The problem is that the slope of your graph, m is only an estimate and any error in that slope will give rise to greater and greater errors in predicted *y* values as *x* values get further from the middle of your graph. The line fitted to data is most reliable at predicting *y* values when *x* is close to the mean, $\bar{X}$ in your experiment. (Note that in this chapter small *x* and small *y* are used when talking about the underlying relationships between variables, but capital *X* and *Y* are used of actual measurements of *x* and *y*. This convention is intended as a reminder that measurements are subject to statistical error and are estimates.)

Figure 6.10 shows a number of lines fitted to imprecise data each of which is plausible. You might have produced any of these straight lines if you had taken this set of data. You can see clearly that the uncertainty in the slope and intercept of the line that defines the relationship between *x* and *y* gives rise to uncertainty over *y* values predicted. The level of unreliability of predicted *y* values is greater when *x* is far from $\bar{X}$ in the experiment.

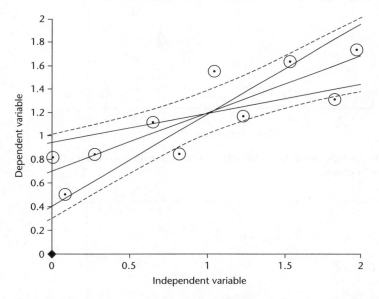

Figure 6.10 The area enclosed with dashed curves indicates confidence limits on *y* values found by interpolation.

Introduction to transformation

Data are often transformed before they are plotted for graphical line fitting. Transformation is covered in more detail after a discussion of statistical methods of line fitting, but here is a quick introduction to the topic. It is generally easier to fit straight lines with a ruler than to fit curves with a flexicurve, and it is often possible to engineer a straight line equation out of a non-linear equation. Let's take one simple example:

$$y = ka^x$$

If you plot a graph of x and y you shouldn't expect a straight-line relationship because this isn't the equation of a straight line. You can get a straight-line relationship if you take the log of both sides of the equation:

$$\log(y) = \log(ka^x) = \log(a) \, x + \log(k)$$

So, if you calculate the logs of the Y data you can plot $\log(Y)$ against X and expect a straight-line relationship with a slope of $\log(a)$ and intercept on the y-axis of $\log(k)$. Transformation means taking either X or Y data and processing them mathematically before fitting your line to the transformed data. A double transform is when you transform both the X and Y data. Transformation affects both the relationship between variables and whether data are homoschedastic or heteroschedastic. Figures 6.11 and 6.12 illustrate the usefulness of transformation.

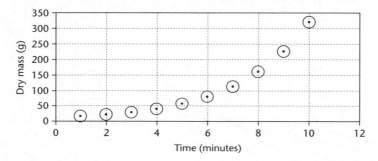

Figure 6.11 Growth rate of bacterial culture. Such a growth is exponential when nutrients are in excess.

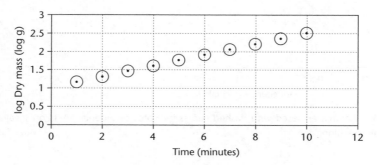

Figure 6.12 Growth rate of bacterial culture – log plot. A log plot of the bacterial growth data gives a straight-line graph.

Correlation

The main statistical method presented in this chapter is regression, but it is appropriate first to talk about correlation. Correlation is not a line-fitting technique but it is concerned with testing the significance of a relationship between two variables. Correlation can be applied to data that are bivariate or univariate, linear or non-linear, homoschedastic or heteroschedastic. Although all these types of data can be analysed, the significance test is looking for the amount of linear nature in the relationship between *x* and *y*. The null hypothesis for this significance test is that there is no linear relationship at all between *x* and *y* values. If the test disproves the null hypothesis then that doesn't necessarily mean the relationship is perfectly linear; it means simply that there is some sort of relationship with a linear component.

Correlation does *not* provide estimates of the constants of an equation that relates *x* to *y*. So what does it give? Correlation calculates a statistic, *r*, called the **product moment correlation coefficient.** If the data points all lie on a straight line that slopes upwards, *r* will be 1.0. The value of *r* will still be 1.0 regardless of how steep the slope is just as long as the data points perfectly line up on an upward-sloping line. If the line slopes downwards and the points are perfectly lined up on it *r* will be –1.0. As the data points are scattered more and more, *r* will tend to move nearer zero. See Figures 6.13 and 6.14.

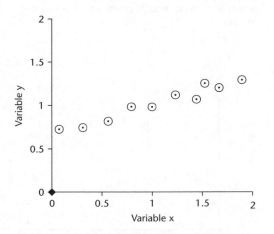 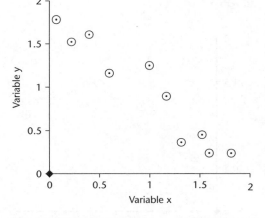

Figure 6.13 Correlation of two variables. These points lie close to a straight line that slopes upwards, so *r* will be close to 1.

Figure 6.14 Correlation of two variables. These points lie near a line that slopes downwards. Although the line is steep, the steepness does not affect *r*.

When there is no relationship between *x* and *y* at all you will tend to see either an elliptical cloud of data points (if the data are bivariate) or a horizontal band of data points (if the data are univariate). When the shape of the cloud of data points is symmetrical about either a horizontal line or a vertical line, or both, there will not be a significant relationship (Figures 6.15 and 6.16).

What about non-linear relationships? If the data points line up on a perfect curve, correlation may or may not detect a relationship depending on the shape of the curve. Figures 6.17 and 6.18 show two curves for comparison.

Why bother with correlation when a statistical line-fitting method (regression) would be capable of finding the constants that define the relationship? One import-ant reason is that correlation is valid for bivariate data as well as univariate data

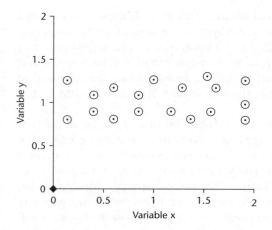

Figure 6.15 Correlation of two variables. These points lie in a horizontal band, so r will be close to zero.

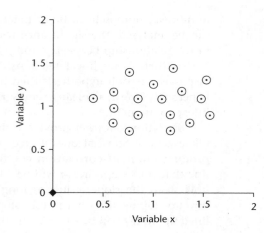

Figure 6.16 Correlation of two variables. These points lie in an elliptiocal cloud, so r will be close to zero.

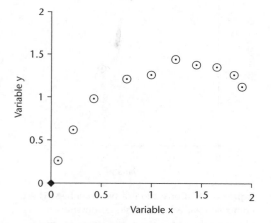

Figure 6.17 Correlation of two variables. This curve has an upward-sloping trend, so r will be positive. Because the data deviate from a straight line, r will be much lower than 1.

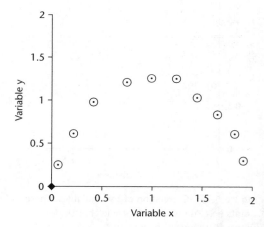

Figure 6.18 Correlation of two variables. This curve is symmetrical and r will be near to zero.

whereas regression is only valid for univariate data and is not valid for bivariate data. In the introduction to this chapter, five example scenarios were presented. Example 5, studying the relationship between the surface area of a human lung and its length, is a case where regression would not be valid but correlation would be. Correlation is also useful when the experimenter simply doesn't know if there is a relationship between two variables and doesn't know the equation that might define that relationship. If you plot your data and the plot looks a little like a random cloud of points, you might detect a significant relationship that is not obvious from a visual check. Linear regression will always put a line through a set of data even if the plot is completely random, so a significance test can be more useful.

Computing r is best done with suitable computer software, although it isn't difficult to do on paper with a pocket calculator. As with standard deviation, there is an equation for r that best shows its purpose and an alternative rearrangement that is easier to calculate:

$$r = \frac{\sum\left(\frac{(X-\bar{X})(Y-\bar{Y})}{s_x \quad s_y}\right)}{N-1} = \frac{\sum XY - \left(\sum X \frac{\sum Y}{N}\right)}{\sqrt{\left(\sum X^2 - \frac{(\sum X)^2}{N}\right)\left(\sum Y^2 - \frac{(\sum Y)^2}{N}\right)}}$$

(In this equation s_x is the sample standard deviation of the X data and s_y is the sample standard deviation in the Y data.)

Finding the product moment correlation coefficient r is just the first step because a significance test needs to involve a probability. Actually, the correlation significance test is a type of t test, which you will be familiar with from Chapter 5. Here are the remaining calculations:

d.f. $= N - 2$

$$SE = \sqrt{\frac{1 - r^2}{N-2}} \quad t = \frac{r}{SE}$$

If you calculate the previous figures by hand you will still have to rely on a computer (or a statistical table) to find p from t at the appropriate degrees of freedom.

Introduction to regression (least squares methods)

Regression is a statistical method that analyses data to determine the constants that define the relationship between variables. There are several different types of regression and the choice of method depends on the type of data to be analysed. There are two assumptions that all the methods make:

Universal assumptions

1. *All but one variable are determined with perfect precision.*

2. *The dependent variable (or its transform) is governed by the normal distribution.*

If we ignore multidimensional data, these reduce to the single assumption that you must have an independent variable that is determined with perfect precision but the dependent variable may be imprecise. In practice, regression is often applied to data where the independent variable is subject to imprecision and to data where there isn't an independent variable. One should be aware of the results of regression if you break this rule, but at least choose the variable with the best precision for your independent variable. Having determined that the data pass the assumption above, you can choose one of three available methods to apply. Your choice will depend on other properties of the data. Each method (listed below) has its own assumptions.

Method 1 – Linear regression (non-weighted)

Assumptions

1. The data are univariate.
2. There is a linear relationship between the variables *or* the data can be transformed to make a linear relationship between the transformed variables.
3. The dependent variable (after any transformation) is homoschedastic.

Requirements

The data only are required.

Difficulty

The method can be carried out fairly easily on paper with the aid of an ordinary pocket calculator. (The calculations are explained in the last section of this chapter.) It is very easy if you have a more modern statistical calculator. Spreadsheet or specialist statistical software can be used.

Method 2 – Non-linear regression (non-weighted)

Assumptions

1. The data are univariate.
2. The dependent variable is homoschedastic or can be made homoschedastic by transforming the data.

Requirements

1. The data.
2. The equation that describes the relationship between the variables.
3. A first guess at each of the constants in the equation.

Difficulty

The method is practically impossible to perform on paper because it is an iterative method. Only very sophisticated calculators will be capable of performing non-linear regression. Spreadsheet software does not normally have facilities for non-linear regression but add-ins might be available to add that functionality or there may be a short list of predefined equations available. The most popular statistical software will be capable of performing non-linear regression and will allow you to specify your own equation of the relationship you want to analyse.

Method 3 – Weighted linear or non linear regression

Assumptions

The data are univariate.

Requirements

1. The data.
2. Either replicate measurements at each value of X, or your own estimate of the precision for each setting of the independent variable.
3. The equation that describes the relationship between the variables.
4. A first guess at each of the constants in the equation.

Difficulty

This is a more sophisticated method than the non-weighted methods. Some statistical software will be capable of performing weighted regressions but you may have to check its functionality before purchasing it. Actually, weighted regression is not often necessary.

Line-fitting flowchart

You would probably benefit from studying the whole of the chapter before you use this flowchart in earnest, but it is worth giving it some study now.

Is least squares suitable?

Before you start with the flowchart you need to decide whether least squares line fitting is suitable by answering three preliminary questions:

Is the independent variable measured with perfect precision?

No – Then least squares regression is not valid. *But*, you may continue if you simply need a plausible line through the data but you should not make conclusions about the constants produced by the regression. In particular, beware of transforming the independent variable as that may exaggerate the imprecision.

Do you know the equation that describes the relationship between the variables?

No – Then least squares regression is not valid. *But*, there are some general-purpose equations (for example, a polynomial) that can be used to put a plausible line through the data. You should not make conclusions about the constants produced by the regression. You *must* not extrapolate the curve because the constants may not have any real meaning for the system you are studying and the line may deviate radically up or down beyond the region set by your data. A polynomial takes the form:

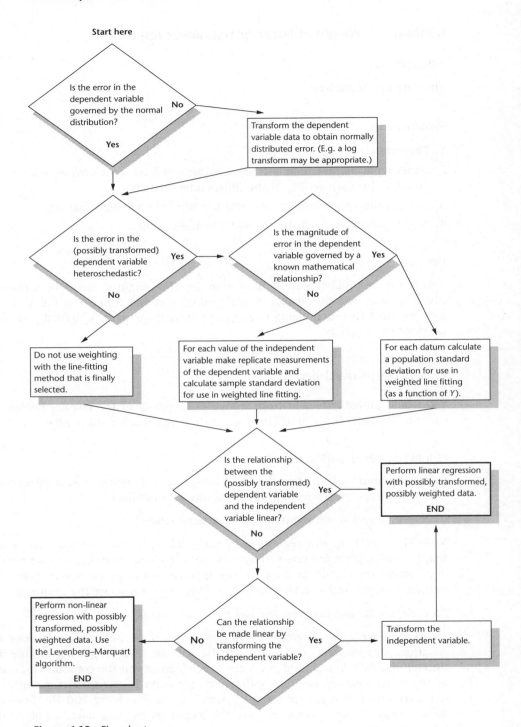

Start here

Is the error in the dependent variable governed by the normal distribution?

No

Yes

Transform the dependent variable data to obtain normally distributed error. (E.g. a log transform may be appropriate.)

Is the error in the (possibly transformed) dependent variable heteroschedastic?

Yes

No

Is the magnitude of error in the dependent variable governed by a known mathematical relationship?

Yes

No

Do not use weighting with the line-fitting method that is finally selected.

For each value of the independent variable make replicate measurements of the dependent variable and calculate sample standard deviation for use in weighted line fitting.

For each datum calculate a population standard deviation for use in weighted line fitting (as a function of Y).

Is the relationship between the (possibly transformed) dependent variable and the independent variable linear?

Yes

No

Perform linear regression with possibly transformed, possibly weighted data.

END

Perform non-linear regression with possibly transformed, possibly weighted data. Use the Levenberg–Marquart algorithm.

END

No

Can the relationship be made linear by transforming the independent variable?

Yes

Transform the independent variable.

Figure 6.19 Flowchart.

$$y = m_0 + m_1x + m_2x^2 + m_3x^3 + m_4x^4 + m_5x^5 + \ldots$$

The number of terms you need depends on the type of curve you have.

Do you have suitable computer software or a suitable pocket calculator?

No – Then you should consider a graphical method of line fitting instead.

Deciding on method of least squares line fitting

The flowchart in Figure 6.19 assumes that you have computer software capable of all the types of least squares line fitting mentioned.

Residuals

Let's look now at some of the mathematics that lies behind regression. Residuals are at the heart of all methods of least squares regression, and residual plots of your data can help you confirm whether the appropriate method of regression was used. Thankfully, residual plots are easy to understand. The residuals can be calculated after you have fitted a line to your data. For a given X datum the residual is the difference between the measured Y datum and the value predicted by the line, denoted $\hat{Y}$. This is illustrated in Figure 6.20. Some residuals will be positive and others will be negative.

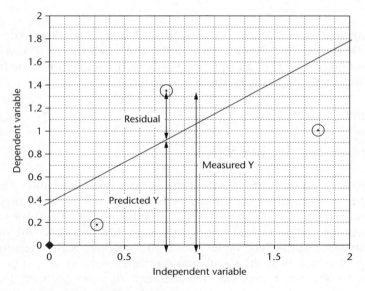

Figure 6.20 Calculating residuals.

It's worth looking at the maths that explains why a measured value might not lie on the line $y = mx + c$ but y is subject to imprecision, so that $Y = y + R$. This gives:

$$Y = mx + c + R$$

Your estimates of m and c will not be perfect because they are estimated from imprecise data. So if M is an estimate of m and C is an estimate of c then, when you predict a value $\hat{Y}$, that value will be given by:

$$\hat{Y} = Mx + C$$

$\hat{Y}$ is an estimate of y based on whatever line-fitting method was applied. The residual checks, for an individual datum, how well that data point corresponds with the line you fitted because:

$$Y - \hat{Y} = (mx + c + R) - (Mx + C)$$

The first term in parentheses corresponds to what actually happened in the experimental process and the second term to a test of your fitted line. The residual can differ from zero for three reasons:

1. The random error on this datum was non-zero.
2. The estimated constants in the equation (one or both) are poor.
3. The equation of the line fitted to the data is not the correct one.

Residuals can be calculated when you use a graphical method of line fitting and when line fitting is done statistically.

A measure of how well the line fits with the data is the sum of squared residuals:

$$\sum (Y - \hat{Y})^2$$

It is this term that gives a name to the method of line fitting described in this chapter – least squares line fitting – because the aim of all the variations of the method is to find constants of an equation that give the minimum value for this statistic. In weighted line-fitting methods a variation is used:

Systematically weighted $\chi^2 = \sum \left(\dfrac{Y - \hat{Y}}{\sigma} \right)^2$

Weighted from data $\chi^2 = \sum \left(\dfrac{Y - \hat{Y}}{s} \right)^2$

These χ^2 statistics are a little different from the one described in Chapter 2 as they aren't found from frequencies, but what they have in common is that they measure scaled differences between observed values and expected values. For each of these a standard deviation is used to weight the residual and they will vary for measurements taken at different values of x. If the precision of the measurement of y is well understood then the population standard deviation, σ, can be calculated, but as an alternative, replicate measurements of y can be made for each value of x and a sample standard deviation, s used.

The actual values of s or σ are not important so long as their relative magnitudes are in the correct proportion. (For example, if every s were doubled this would increase the χ^2 statistic but the same line would be fitted to the data, simply with a bigger final χ^2 value.) So, in some computer programs a non-weighted line-fitting session is started simply by setting all the standard deviations to 1.0. It doesn't matter if that is the wrong standard deviation, so long as the real standard deviation is also constant across all values. That means that a χ^2 statistic is still calculated for unweighted line fitting but it is equivalent to the sum of squared residuals.

The role of residuals in regression

Non-weighted, non-linear regression puts a line through the data using the first-guess constants that you provide. It then finds the sum of the squared residuals (same as χ^2 with s set to 1) as a measure of how good the fit is. (This part of the method could be done easily with a pocket calculator.) Some clever maths is done to determine which constants should be raised and which lowered to make the fit better, and all the constants are adjusted. The constants are iteratively adjusted until the sum of the squared residuals is minimized. It's this repetitive element of the method that makes it practically impossible without a computer.

Weighted, non-linear regression is performed in basically the same way except that χ^2 is calculated by weighting each squared residual by s^{-2}. The weighting factors can be provided by you or, if you provide replicate measurements, the computer software will compute its own weighting factors.

Non-weighted, linear regression is a special case because the least squares line can be found from the data alone, without first guessing constants. It automatically provides a line that gives the minimal sum of squared residuals.

Weighted, linear regression is also possible. The calculation might be a simple variation on the non-weighted version but in practice those with access to computer software will tend to use the iterative software and select the equation of a straight line from the menu of equations.

Residual plots

It is often worthwhile making a residual plot after performing regression. For a residual plot you calculate the residual, $(Y - \hat{Y})$ for each datum, using your equation and the estimated constants determined by the regression to determine $\hat{Y}$. The plot is the residual against the X datum. You can use the residual plot to check the assumptions you made about the data. You can make a visual check to see if the data were homoschedastic and if the data fitted the equation you used for the regression. For an example see Figure 6.21.

The data points appear to be randomly spread around the chart, which is what you would expect. If some pattern or trend were observable this might indicate that false assumptions had been made about the data and linear regression is unsuitable. Two examples of problem data follow.

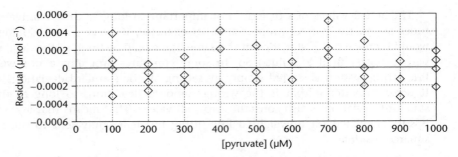

Figure 6.21 LDH reaction kinetics – residual plot. This computer-generated graph uses diamonds for the data points because points with circles are rarely available in computer software.

Data are linear but heteroschedastic

Data, shown in Figure 6.22, were analysed.

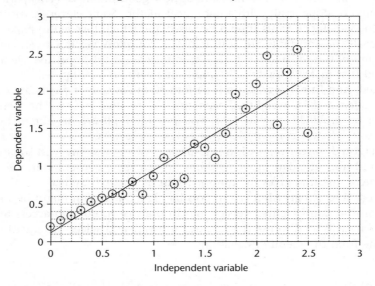

Figure 6.22 Heteroschedastic dependent variable.

The residual plot in Figure 6.23 reveals increasing scatter of Y values as X values increase. This means that a key assumption about the data is false and linear regression is not a valid test.

Data are homoschedastic but not linear

Figure 6.24 shows more data with a straight line fitted.

The residual plot in Figure 6.25 can also highlight non-linearity. The residuals for these data are mostly negative at either end of the chart and positive in the middle. This shows that the data are probably not linear and a linear regression is not a valid test because of this.

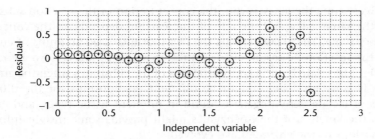

Figure 6.23 Heteroschedastic – residual plot.

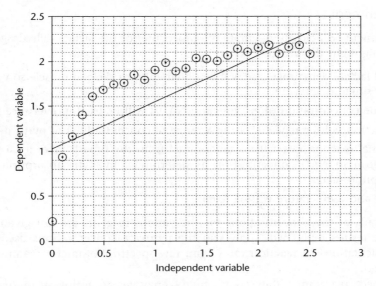

Figure 6.24 Non-linear relationship.

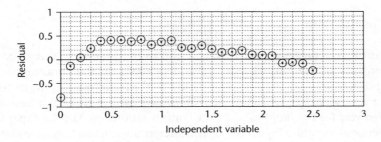

Figure 6.25 Non-linear relationship – residual plot.

(**Case Study**) Lactate dehydrogenase

For the purposes of this example we will look at the reaction kinetics of the enzyme lactate dehydrogenase. The assay for this enzyme was described in Chapter 3 and you will remember that the assay involves looking at the rate of the reaction it catalyses. Every time the assay is run, the concentration of the reaction substrates is fixed at the same value and the rate of the reaction depends on the concentration of the enzyme. The assay exploits the fact that increased amounts of enzyme will make the reaction go faster, but it would be useful to know how the rate of reaction might be related to the concentrations of the substrates when the amount of enzyme is constant. There might be different relationships for the different isoenzymes of lactate dehydrogenase found in different tissue types, and experiments into the kinetics of the isoenzymes might provide some insight into why different versions of the enzyme have evolved.

The experiment

The experiment will closely resemble the procedure for the lactate dehydrogenase assay. To highlight the differences:

1. Purified lactate dehydrogenase will be obtained from skeletal muscle so we can study a single isoenzyme.
2. The same concentration of enzyme and NADH will be used in every reaction.
3. The reaction will be repeated using a range of concentrations of pyruvate.
4. The initial reaction rate will be determined from the time it takes for a fixed, small drop in NAD concentration as determined by light absorption in a spectrophotometer.
5. The pH and temperature will be fixed.

The details of the concentrations for the various reagents are not given here except the results tables will show you the pyruvate concentrations used. The routine calculations of reaction rates from raw spectrophotometer readings are also omitted.

We expect reactions catalysed by many enzymes to follow a predictable relationship;

Actual relationship: $v_0 = \dfrac{V_{max}[\text{pyruvate}]}{K_m + [\text{pyruvate}]}$

The constant K_m is a property of the enzyme and the constant V_{max} is derived from another property of the enzyme but is also dependent on the concentrations of enzyme and NADH used in the experiment. Not all enzymes conform to this pattern but the isoenzyme of lactate dehydrogenase used in this study behaves in the expected way for the range of pyruvate concentrations used in the experiment. The independent variable, [pyruvate], can be determined using simple volumetric methods and is in effect measured with perfect precision. The reaction velocity is subject to imprecision, but if the experiment is carried out effectively the only significant contributor to imprecision will be the spectrophotometer. Since every run of the reaction will involve absorbance readings on the spectrophotometer within the same range, it is reasonable to assume that the precision will be the same for low- and high-reaction velocities, i.e. the dependent variable is homoschedastic.

For the next part of this chapter you should imagine that the relationship between v_0 and [pyruvate] isn't known and you need to analyse the experimental results to determine the relationship. In the next section we test out a linear relationship:

Hypothetical (false) relationship: $v_0 = m[\text{pyruvate}] + c$

Application of linear regression

Linear regression can be used simply to determine values for the coefficients of a straight line that gives the best fit to the data. It is also possible to apply a variety of significance tests alongside these core calculations. If you use a computer to perform your linear regression it's likely that an array of significance tests will be applied whether you ask for them or not, so in this section we look at a worked example and explain the various elements of the computer output. Here are the assumptions we are going to make in order to apply linear regression:

1. The dilutions of pyruvate are achieved with simple volumetric methods so there is no error at all in the determination of concentrations. (A valid assumption in this case.)

2. The precision of measurements of initial reaction rate may suffer imprecision but the amount of imprecision is constant for experiments at high and low [pyruvate] and is normally distributed. (Also a valid assumption in this case.)

3. The relationship between initial reaction rate and [pyruvate] is linear. (This is a working assumption because this is something we would like to find out. In other words, we are trying out a linear relationship to see how well it fits the data and we can reject the assumption if it isn't consistent with the data.)

So in an experiment, initial reaction rate was determined for reaction mixtures with differing substrate concentration. Results are presented in Table 6.1.

Table 6.1	
[pyruvate] **μM**	**Reaction rate** **(μmol s^{-1})**
25	0.0032
50	0.0056
75	0.0075
100	0.0105
125	0.0120
150	0.0139
175	0.0150
200	0.0167
225	0.0178
250	0.0187

The regression option of the Analysis Tools command in Excel produces a bewildering amount of statistics (see Table 6.2).

Table 6.2

	A	B	C	D	E	F	G
1	SUMMARY OUTPUT						
2	*Regression statistics*						
3	Multiple R	0.988793569					
4	R Square	0.977712722					
5	Adjusted R	0.974926813					
6	Square						
7	Standard Error	0.000841283					
8	Observations	10					
9							
10	ANOVA						
11		*df*	*SS*	*MS*	*F*	*Significance F*	
12	Regression	1	0.000248387	0.000248387	350.9491779	6.80762E-08	
13	Residual	8	5.66206E-06	7.07758E-07			
14	Total	9	0.000254049				
15							
16		*Coefficients*	*SE*	*t Stat*	*P-value*	*Lower 95%*	*Upper 95%*
17	Intercept	0.002546667	0.000574706	4.431252015	0.00219346	0.001221392	0.003871942
18	X Variable 1	6.94061E-05	3.70489E-06	18.73363761	6.80762E-08	6.08626E-05	7.79496E-05

Let's skip over the statistics to the answer. The rate is labelled 'X Variable 1' in the bottom row of the output. (Excel can perform multidimensional linear regression so the multiple independent variables are labelled X Variable 1, X Variable 2, X Variable 3 etc., each having its own coefficient and associated statistics. With just one independent variable the label X Variable 1 is used in place of m.) So the coefficient m in our putative linear relationship is 6.94×10^{-5} μmol s^{-1} μM^{-1}. However, if you look along the row you will see 95% confidence limits on the rate – there is a 0.95 probability that the rate lies between 6.09×10^{-5} and 7.79×10^{-5} μmol s^{-1} μM^{-1}. Quite a wide range – when you fit a line to data using a ruler you tend to overestimate the reliability of the slope but the regression quantifies the uncertainty. The coefficient c in our equation of a straight line is 0.0025 μmol s^{-1} with upper and lower 95% confidence limits of 0.0039 μmol s^{-1} and 0.00 2 μmol s^{-1}. Now go back through some of the key statistics printed in Table 6.2.

Product moment correlation coefficient – (Multiple) R

R = 0.988793569. The computer software does correlation calculations in addition to the regression you ask for. This statistic is r, the measure of how closely initial reaction velocity and pyruvate concentration vary together. If you wanted you could make a few more calculations and do a significance test on r to check that v_0 and [pyruvate] are related. Just follow the maths described in the section on correlation above. You should find that the AnoVa table is actually a clearer indication of the significance of the relationship but r is often quoted in literature, as it's simpler to calculate on paper. There are some other figures connected with r in the same table, which we will skip over.

Analysis of variance

This part of Table 6.2 should be familiar from Chapter 5. Analysis of variance can be applied to regression data to measure the significance of the relationship between the independent and dependent variables. As usual, sources of variation are partitioned. The AnoVa takes the intercept and X Variable 1 of the regression and predicts the reaction velocities that should have been obtained at the given pyruvate concentrations if there was a perfect linear relationship. For each [pyruvate] value a residual is also calculated. The residual is the deviation of the measured v_0 (i.e. Y) from the predicted v_0 (i.e. $\hat{Y}$). There is variation due to the regression and there is variation due to the random deviations from predicted values – the residual variation. If there is a strong relationship between reaction velocity and pyruvate concentration the regression mean squares will be much higher than the residual mean squares. An f statistic is measured in exactly the same way as for an ordinary single factor AnoVa (as described in Chapter 5).

In our case $f = 350.9491779$ which gives $p = 6.80762 \times 10^{-8}$. So the probability that the data points lined up by accident alone is virtually zero. You could have guessed that from the graph of our data (which follows later), but when there is greater random variation in the dependent variable a visual check of a graph isn't good enough and a statistical test is needed. The human brain is very good at making patterns out of random dots, so there is a tendency to see a line where there isn't one – the AnoVa gives you a systematic method of checking for significance.

The coefficients

The last part of Table 6.2 gives you the result of the regression and performs some statistical analysis using the t distribution. The two coefficients m and c are treated separately. In the printout the coefficient we are calling m is referred to as X Variable 1 and the coefficient we refer to as c is referred to as Intercept. Two operations are performed on each coefficient: the significance of the deviation from zero is determined and confidence limits are attached. Take the X Variable 1 row cell by cell, as follows.

Standard error

$$\text{SE} = 3.70489 \times 10^{-6} \; \mu\text{mol s}^{-1} \, \mu\text{M}^{-1}$$

This is just like the standard error discussed in Chapter 4, except that it is a measure of the precision of the coefficient rather than the precision of the measurements themselves.

The t statistic

$$t = 18.73363761$$

This is just like the t statistic discussed in Chapter 5. It is the standardized deviation of the coefficient from zero, so it is a statistical measure of how far the coefficient m is from 0.0. Like the usual t value it is the deviation divided by the standard error.

The P-value

$$p = 6.80762 \times 10^{-5}$$

The t distribution is used to calculate a probability from the t statistic. This is the probability that the measured m deviates from zero only because of random variation in the data points. Since p is vanishingly small there must be a very significant m. (Significant doesn't mean that m is big – it just means that m is definitely not zero.)

Lower 95% and Upper 95%

$$\text{Lower 95\%} = 6.09 \times 10^{-5} \; \mu\text{mol s}^{-1} \, \mu\text{M}^{-1}$$

$$\text{Upper 95\%} = 7.79 \times 10^{-5} \; \mu\text{mol s}^{-1} \, \mu\text{M}^{-1}$$

These are the 95% confidence limits on the rate of growth calculated from its standard error.

A similar interpretation applies to the coefficient c, labelled Intercept in the table of statistics.

Making a report

How does all this translate into plain English for a report? Boiling down all this information, here is one interpretation. Analysis of variance applied to the regression statistics for the data gave an f statistic of 350.9491779 at 1 and 8 degrees of freedom. The probability that the observed regression occurred because of chance variation in measured initial reaction velocities is 6.80762×10^{-8} and so it must be concluded that there is a highly significant regression between substrate concentration and initial reaction velocity. The regression analysis gives a slope of 6.94×10^{-5} µmol s^{-1} µM^{-1}. There is 95% confidence that the true slope lies between 6.09×10^{-5} and 7.79×10^{-5} µmol s^{-1} µM^{-1}. The slope is highly significantly non-zero as the probability of its occurrence by chance variation in reaction rates is 6.80762×10^{-5}. The regression analysis gives an intercept of 0.0025 µmol s^{-1}. There is 95% confidence that the true intercept lies between 0.0012 and 0.0039 µmol s^{-1}. The intercept is highly significantly non-zero as the probability of its occurrence by chance variation in initial reaction velocities is 0.0022.

There is an omission from this report: it doesn't address the doubtful assumption that the relationship in linear.

Just because we found a straight line to fit the data with a significant positive slope, that doesn't confirm that the data are definitely linear. In fact there is a glaring inconsistency which is hidden somewhat by the statistical language. The report demonstrates that the intercept is significantly non-zero, and that needs to be related back to the system we are studying. It means that the linear regression is predicting that there will be a significant reaction rate when there is no substrate at all, and that is plainly impossible. That isn't a direct test for non-linearity but it is an important piece of evidence against the linear relationship hypothesis.

A residual plot

Despite all the sophisticated statistical methods a simple graphical representation of the data can reveal problems that you might not otherwise have seen. There are two useful plots: a plot of [pyruvate] and v_0 (Figure 6.26) and a then a residual plot (Figure 6.27).

The residual plot seems to show a trend for residuals to be negative at the extreme x values and to be positive in the middle. If that didn't happen by accident then a curve might better fit the data. For a non-scientist the application of linear regression would seem to have been a mistake at this stage but this has been a reasonable step in a scientific method: we don't know the maths of this system so we start with a working assumption (a hypothesis) that can be tested. Disproving the hypothesis takes us further forward in our understanding, just as proving the hypothesis would also take us forward. As things stand, we haven't completely disproved the linearity hypothesis and a systematic test of linearity is needed.

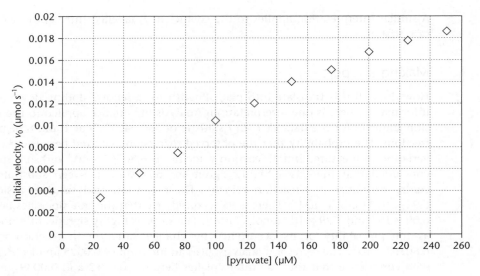

Figure 6.26 LDH reaction kinetics.

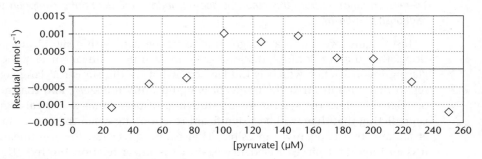

Figure 6.27 LDH reaction kinetics – residual plot.

Testing for significant non-linearity

To carry out the linear regression it was necessary to make the assumption that there was a linear relationship between the independent and dependent variables. Graphical methods have placed doubt on this assumption, and even if they hadn't, we would still want to test the working assumption that the relationship is linear. In this situation it is useful to have a significance test that can tell you if the distribution of data is consistent with that assumption. The null hypothesis in this situation is that residuals from the linear regression are non-zero only because of uniform imprecision in the measurements of y. This null hypothesis can be disproved if there is a trend of different residuals at different values of x. The best way to test this experimentally is to replicate the measurement of initial reaction velocity at each value of pyruvate concentration. In Table 6.3 you can see that the reaction rate experiment was repeated several times at each substrate concentration. This allows for the quantification of precision.

Table 6.3

[pyruvate]	25	50	75	100	125	150	175	200	225	250
Velocities	0.0031	0.0058	0.0081	0.0098	0.0119	0.0134	0.0148	0.0163	0.0174	0.0189
	0.0030	0.0059	0.0079	0.0096	0.0119	0.0136	0.0148	0.0165	0.0176	0.0185
	0.0027	0.0056	0.0082	0.0102	0.0120	0.0133	0.0147	0.0163	0.0178	0.0184
	0.0034	0.0059	0.0079	0.0102	0.0116	0.0136	0.0144	0.0160	0.0178	0.0187

The null hypothesis can then be tested with a type of AnoVa. It's likely that your spreadsheet software will be unable to perform the appropriate statistics (e.g. Excel) but you can get most of the calculations done using automated facilities of the software and then add a few extra calculations to perform the extra significance test. Instructions for doing this are included in Chapter 8 but the mathematics is explained here.

Stage 1 – Regression

The first part of a non-linearity test is to perform a regression to obtain slope and intercept of the best fitting line. This has to be done with replicate measurements at each value of x. Results of this analysis are shown in Table 6.4.

Table 6.4

	A	B	C	D	E	F	G
1		Coefficients	SE	t Stat	P-value	Lower 95%	Upper 95%
2	Intercept	0.002598	0.000232	11.17703	1.42E–13	0.002128	0.003069
3	X Variable 1	6.79E–05	1.5E–06	45.32183	1.09E–34	6.49E–05	7.1E–05

Stage 2 – Standard AnoVa

In the second stage the same data are organized into groups and an AnoVa performed. (This is identical to the AnoVa described in Chapter 5 for block design.) The summary table from the AnoVa, Table 6.5, tells you the mean and variance of each group of four replicates.

The analysis proper, Table 6.6, measures the total amount of variation within the replicates and measures the total amount of variation between group means.

Table 6.5

	A	B	C	D	E
1	SUMMARY				
2	[pyruvate]	Count	Sum	Average	Variance
3	25	4	0.0122	0.00305	8.33E−08
4	50	4	0.0232	0.0058	2E−08
5	75	4	0.0321	0.008025	2.25E−08
6	100	4	0.0398	0.00995	9E−08
7	125	4	0.0474	0.01185	3E−08
8	150	4	0.0539	0.013475	2.25E−08
9	175	4	0.0587	0.014675	3.58E−08
10	200	4	0.0651	0.016275	4.25E−08
11	225	4	0.0706	0.01765	3.67E−08
12	250	4	0.0745	0.018625	4.92E−08

Table 6.6

	A	B	C	D	E	F	G
1	ANOVA						
2	Source of variation	SS	df	MS	F	P-value	F crit
3	Between groups	0.000968	9	0.000108	2486.31	1.45E−40	2.210697
4	Within groups	1.3E−06	30	4.33E−08			
5							
6	Total	0.000969	39				

Stage 3 – Calculating mean squared mean residuals

The third stage is to calculate the mean squared deviation of each group mean from its value predicted by the regression line. In symbols:

$$\frac{(\bar{Y} - \hat{Y})^2}{N}$$

This can be done on a spreadsheet by extending the summary table from the AnoVa as shown in Table 6.7.

Stage 4 – Non-linearity AnoVa

The standard AnoVa measures the variation between the means of groups of data but that variation has two components:

1. Variation due to a linear relationship between the variables x and y, i.e. due to regression.

2. Extra variation due to non-linearity.

The symbols used in the following explanation are presented in Table 6.8.

Table 6.7

	A	B	C	D	E	F	G
1	SUMMARY						
2	Groups	Count	Sum	Average	Variance	$\hat{Y}$	$(\bar{Y}-\hat{Y})^2/N$
3	25	4	0.0122	0.00305	8.33E−08	0.004296	3.88356E−07
4	50	4	0.0232	0.0058	2E−08	0.005994	9.44725E−09
5	75	4	0.0321	0.008025	2.25E−08	0.007692	2.76517E−08
6	100	4	0.0398	0.00995	9E−08	0.00939	7.82728E−08
7	125	4	0.0474	0.01185	3E−08	0.011088	1.44976E−07
8	150	4	0.0539	0.013475	2.25E−08	0.012787	1.18503E−07
9	175	4	0.0587	0.014675	3.58E−08	0.014485	9.06823E−09
10	200	4	0.0651	0.016275	4.25E−08	0.016183	2.13556E−09
11	225	4	0.0706	0.01765	3.67E−08	0.017881	1.32948E−08
12	250	4	0.0745	0.018625	4.92E−08	0.019579	2.27356E−07

Table 6.8

Abbreviation	Explanation	Abbreviation	Explanation
SSG	Sum of squares between groups	MSG	Mean squares between groups
SSR	Sum of squares for regression	MSR	Mean squares for regression
SSNL	Sum of squares for non-linearity	MSNL	Mean squares for non-linearity

$$SSNL = \sum \frac{(\bar{Y}-\hat{Y})^2}{N}$$ (i.e. total up the column in the previous stage)

$$SSR = SSG - SSNL$$

As with any other AnoVa you calculate mean squares by dividing each sum of squares by the appropriate degrees of freedom. Degree of freedom for non-linearity is the number of groups minus 1. Degree of freedom for regression is 1. Now you are in a position to calculate an f statistic that compares variation due to non-linearity with the variation due to imprecision (variation within groups):

$$f = \frac{MSNL}{MSG}$$

When AnoVa is done entirely by the computer p is calculated with no fuss, but now that you've calculated your own f statistic you need to use the appropriate spreadsheet function to calculate p:

=FDIST(f, df1, df2)

This is the probability of the null hypothesis; that the data values deviate from the regression straight line only because of uniform imprecision in the determination of y. Table 6.9 is the original AnoVa table with two extra rows added by hand according to the instructions above.

Table 6.9

	A	B	C	D	E	F	G
1	ANOVA						
2	*Source of variation*	*SS*	*df*	*MS*	*F*	*P-value*	*F crit*
3	Between groups	0.000968	9	0.000108	2486.31	1.45E–40	2.210697
4	Regression	0.000967	1	0.000967			
5	Non-linearity	1.02E–06	9	1.13E–07	2.618011	**0.023033**	
6	Within groups	1.3E–06	30	4.33E–08			
7							
8	Total	0.000969	39				

The probability that the group means deviate from the regression line only because of imprecision in y is 2.3%. This test effectively disproves the hypothesis that there is a linear relationship between initial reaction velocity and substrate concentration. Where do you go from here? We will demonstrate two more methods of linear regression which are improvements on the method presented here. Both of these will have problems of their own. Finally we will apply a non-linear regression.

Transforming data

Transforming data means applying some mathematical function to either the dependent or the independent variables, or both. Common transforms are the inverse or the log of the data. The actual X value or Y value is transformed but the transformed value is used as input to the regression as if it were original data. There are three reasons for transforming your data. In order of importance they are:

1. To compensate for non-normally distributed error in the dependent variable.

2. To avoid the use of weighted regression with heteroschedastic data.

3. To make valid the use of linear regression with non-linear data.

Data should not be transformed to allow linear regression if you know it will result in non-normally distributed error in the dependent variable. That's because all least square line-fitting methods assume normal distribution in Y. There is a long tradition of misusing the transformation of data to make linear relationships at all costs, specifically at the cost of creating or emphasizing the non-normal distribution in the data. This tradition has arisen out of the practical impossibility of performing non-linear regression without a computer. If you simply don't have the facilities to perform non-linear regression it is very tempting to transform it.

Often, transformation will simultaneously make the error normal and also make it homoschedastic but it is not valid to use transforms to give homoschedasticity if it makes the error non-normal. Alternatively, the transform that makes the data normal may make it heteroschedastic. In an experimental situation you may not know if the transform will cause a problem but you may, in some circumstances, hypothesize that data will be homoschedastic after transformation and then test that hypothesis statistically. We will make that working assumption for the enzyme kinetics example and attempt a linear regression of transformed data. Heteroschedastic data will require a weighted regression.

How, then, should the data be transformed? You must take the equation that describes the relationship between substrate concentration and initial reaction velocity and choose a transform that will result in the equation of a straight line. The problem is that we don't know what the appropriate equation is for our enzyme kinetics study. (Or at least we're pretending we don't know.) The graph of the data collected so far doesn't suggest any obvious type of equation but it might be revealing to see what the curve does at higher values of substrate concentration than were used before. Table 6.10 contains data from a repeat experiment with a wider range of substrate concentration. When plotted, we get the graph shown in Figure 6.28.

Table 6.10

[pyruvate]	µmol s^{-1}	[pyruvate]	µmol s^{-1}
100	0.0101	600	0.0283
100	0.01	600	0.0285
100	0.0097	600	0.0283
100	0.0104	600	0.0285
200	0.0165	700	0.0298
200	0.0167	700	0.0298
200	0.0164	700	0.0297
200	0.0166	700	0.0294
300	0.0209	800	0.0312
300	0.0207	800	0.0314
300	0.021	800	0.0313
300	0.0207	800	0.0309
400	0.0238	900	0.0322
400	0.0236	900	0.0324
400	0.0242	900	0.0326
400	0.0242	900	0.0326
500	0.0266	1000	0.0334
500	0.0265	1000	0.0331
500	0.0266	1000	0.033
500	0.0262	1000	0.0332

You can now see clearly that the relationship is non-linear and there seems to be a maximum initial reaction velocity which is approached by the use of higher and higher substrate concentrations. (If you think of enzyme molecules as tiny machines you can imagine that there is a limit to how fast substrate molecules can be processed.) The shape of the graph is recognizable to mathematicians as a rectangular hyperbola and there is a well known form of equation that describes a hyperbola:

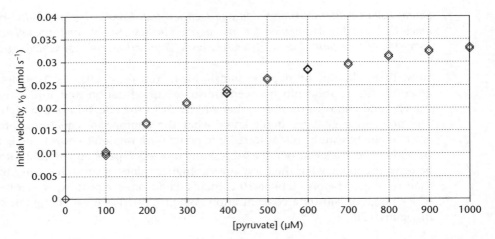

Figure 6.28 LDH reaction kinetics. Replicate measurements are close to each other and some lie on top of others.

$$v_0 = \frac{V_{\max}[\text{pyruvate}]}{K_m + [\text{pyruvate}]}$$

The coefficients in this equation are the conventional ones used by biochemists. The coefficient that defines how high the hyperbola goes is $V_{\max}$, the maximum reaction velocity. The other coefficient, K_m is called the Michaelis–Menton constant and is named after the two researchers who worked out this equation. (These constants are related to the mechanism by which enzymes operate. In our experiment we ought to use the phrases 'apparent K_m' and 'apparent $V_{\max}$' to remind ourselves that the measured values may be affected by our choice of experimental conditions.) The problem now is to find a transform of this equation that gives the equation of a straight line. There is more than one solution but we can start with the most traditional solution. Take the reciprocal of both sides of the equation and rearrange. (If you are rusty on algebra and the following doesn't make sense, it might help to refer back to the section on algebra of exponents in Chapter 4. A method using exponent algebra has been chosen to avoid introducing any new algebra but you may use your own method.)

First a representation of the Michaelis–Menton equation using exponents:

$$v_0 = V_{\max}[\text{pyruvate}](K_m + [\text{pyruvate}])^{-1}$$

Take the reciprocal of both sides:

$$v_0^{-1} = (V_{\max}[\text{pyruvate}](K_m + [\text{pyruvate}])^{-1})^{-1}$$

If several terms are multiplied and the product raised to a power of n, that is the same as raising each term to the power of n first and then finding the product. Hence:

$$v_0^{-1} = V_{\max}^{-1}[\text{pyruvate}]^{-1}\{(K_m + [\text{pyruvate}])^{-1}\}^{-1}$$

A term raised to a power of n and the result raised to a power of m is the same as the term raised to a power of the product of n and m, and since $-1 \times -1 = 1$:

$$v_0^{-1} = V_{max}^{-1}[\text{pyruvate}]^{-1}(K_m + [\text{pyruvate}])$$

This can be rearranged as:

$$v_0^{-1} = V_{max}^{-1}[\text{pyruvate}]^{-1}K_m + V_{max}^{-1}[\text{pyruvate}]^{-1}[\text{pyruvate}]$$

And then simplified to:

$$v_0^{-1} = V_{max}^{-1}K_m[\text{pyruvate}]^{-1} + V_{max}^{-1}$$

Using the more familiar division notation instead of exponents:

$$\frac{1}{v_0} = \frac{K_m}{V_{max}}\frac{1}{[\text{pyruvate}]} + \frac{1}{V_{max}}$$

If you think of $1/v_0$ as y and $1/[\text{pyruvate}]$ as x you have the equation of a straight line with a slope of K_m/V_{max} and intercept on the y-axis of $1/V_{max}$. The plot of reciprocal initial reaction velocity against reciprocal of substrate concentration is known as a Lineweaver–Burke plot after the researchers who devised it. Table 6.11 presents the transformed data and Figure 6.29 is the graph.

Table 6.11

1/[pyr]	0.01	0.005	0.00333	0.0025	0.002
1/v_0	99.0099	60.6061	47.8469	42.0168	37.594
	100	59.8802	48.3092	42.3729	37.7358
	103.093	60.9756	47.619	41.3223	37.594
	96.1538	60.241	48.3092	41.3223	38.1679

1/[pyr]	0.00167	0.00143	0.00125	0.00111	0.001
1/v_0	35.3357	33.557	32.0513	31.0559	29.9401
	35.0877	33.557	31.8471	30.8642	30.2115
	35.3357	33.67	31.9489	30.6748	30.303
	35.0877	34.0136	32.3625	30.6748	30.1205

Performing the linear regression

The application of linear regression makes a number of assumptions:

1. The relationship between the inverse of initial reaction velocity and the inverse of pyruvate concentration is linear. (Actually a valid assumption – but we're still pretending we don't know that yet.)

2. Values for the inverse of pyruvate concentration are found with perfect precision. If we can assume the untransformed pyruvate concentrations are found with perfect precision, transformation isn't going to introduce any new imprecision, so this seems a reasonable assumption. If there is a tiny amount of imprecision in the untransformed data the transform will emphasize it at one extreme of the graph so we need to be careful of this assumption.

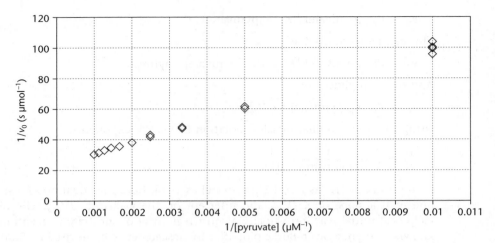

Figure 6.29 LDH reaction kinetics – double reciprocal. It is difficult to judge the amount of scatter on this graph.

3. Values for the inverse of reaction rate are assumed to be found with equal precision for high and low values of pyruvate concentration. This is a dubious assumption because the untransformed data are homoschedastic and the transform will exaggerate imprecision for reaction rates measured at low pyruvate concentrations. We test this assumption statistically at the end of this section.

The linear regression is performed on the transformed data in the conventional manner but there are some additional concerns beyond finding the coefficients that define the relationship. We need to establish whether the transformed data are significantly non-linear to test our assumption that the curve of the untransformed variables really is a hyperbola. We also need to find out whether the transformed reaction velocities are significantly heteroschedastic, because if they are, this will cast doubt on the validity of the coefficients that were determined. (We are still pretending we don't already know the answers to these questions, but just in case you have forgotten, the transformed data really are linear so we shouldn't find significant non-linearity. The transformed reaction velocities, however, *are* heteroschedastic.)

The coefficients

With the transformed data, finding the coefficient means first applying linear regression in the standard way but using the transformed data in place of X and Y data. The computer printout is shown in Table 6.12.

Table 6.12

	A	B	C	D	E	F	G
1		Coefficients	SE	t Stat	P-value	Lower 95%	Upper 95%
2	Intercept	22.40057	0.205589	108.9582	4.82E–49	21.98438	22.81676
3	X Variable 1	7697.645	52.22346	147.3982	5.11E–54	7591.924	7803.366

Getting the coefficients from the slope and intercept means doing a little maths. The estimate of V_{max} is the reciprocal of the intercept:

$V_{max} = 0.044641726$ µmol s^{-1}

The estimate of K_m is found from the slope of the regression and V_{max}:

$K_m = 343.6361405$ µM

Testing linearity

The next step is to check for significant non-linearity in the transformed data. If there is significant non-linearity we may have to reject the hyperbola relationship that we hypothesized. This significance test is exactly the same procedure as described earlier. The results from the analysis are in Table 6.13.

Table 6.13

	A	B	C	D	E	F	G
1	ANOVA						
2	*Source of variation*	*SS*	*df*	*MS*	*F*	*P-value*	*F crit*
3	Between groups	16400.15	9	1822.239	2011.926	3.44E–39	2.210697
4	Regression	16400.06	1	16400.06			
5	Non-linearity	0.094392	9	0.010488	0.01158	1	
6	Within groups	27.17157	30	0.905719			
7							
8	Total	16427.32	39				

The probability of the null hypothesis is so high that the computer software can't distinguish it from 1.0, the maximum, so we can accept that the transformed data were linear and it supports our hyperbola as a reasonable description of the relationship between the variables. (It's important to note that the result doesn't make it certain that we have a hyperbola, particularly if the underlying system deviates from a hyperbola for [pyruvate] values higher than we used in our experiment.)

Testing homoschedasticity

This also requires replicate measurements so we can analyse the same data from the previous experiment and in fact use some of the figures calculated there. Because replicates were made we can estimate precision at each value of X. One measure of precision is variance and this is convenient because the previous AnoVa provided exactly that in the summary statistics table. A rigorous significance test would take into account the variance of every block, but in the interests of simplicity we will compare the block with the highest variance against the block with the lowest variance. In general, an analysis of variance compares two variance figures by calculating their ratio: that's what the f statistic is. So the significance test we need, called an f_{max} test, is actually about the easiest AnoVa you will ever apply:

$$f = \frac{s_{max}^2}{s_{min}^2}$$

There is a catch. All the previous AnoVa described in this book have been one-tailed tests because you were looking for variation *higher than* the variation due to imprecision. In this test we need a two-tailed test because if the null hypothesis really is true the block with the higher variance might just as easily have been the block with the lower variance. So, if your computer gives you a one-tailed probability you will need to multiply by two. For example:

=FDIST(f,N-1,2)*2

This simple analysis only works if the selected blocks have the same number of replicate measurements. N in the spreadsheet formula is the number of replicates in one block. The results of the analysis are presented in Table 6.14.

Table 6.14

	A	B	C	D
	S_{max}^2	S_{min}^2	F_{max}	P
1				
2	8.192861	0.020496	399.7233	0.004993

(Note that this AnoVa was comparing variance of groups of replicate measurements whereas the non-linearity AnoVa was looking at variance of residuals.) This analysis reveals that the transformed data are very highly significantly heteroschedastic. The residual plot in Figure 6.30 gives a graphical view of the problem we have.

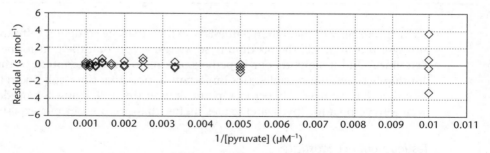

Figure 6.30 LDH reaction kinetics – double reciprocal residual plot. This residual plot shows the heteroschedastic nature of the transformed data. This was much less obvious in Figure 6.29.

This is telling us that the double reciprocal plot is invalid as a method of determining V_{max} and K_m. Is there an alternative transform that will allow linear regression? There is another possible transform we could consider. Take the Michaelis–Menton equation, multiply both sides by V_{max} and then rearrange to obtain:

$$v_0 = -K_m \frac{v_0}{[\text{pyruvate}]} + V_{max}$$

This does allow a straight-line plot of v_0 against $v_0/[\text{pyruvate}]$ with untransformed reaction velocities on the y-axis. It solves one problem but generates another. The dependent variable (on the y-axis) is not transformed but we now have a supposedly independent variable (v_0/[pyruvate]) that is now subject to the imprecision from the reaction velocity measurements. This transform, an Eadie–Hofstede plot has one small advantage: non-linearity is magnified if it exists. Really, a different approach is needed and we still have two possible lines of attack:

1. Weighted linear regression. The idea is to transform the data to get a linear relationship but then to compensate for the heteroschedastic nature by weighting the squared residuals appropriately.

2. Non-weighted, non-linear regression. Here the idea is to fit a curve to the original data, without transform. Because the initial reaction velocities are homoschedastic, weighting is not needed.

We will try both of these approaches and then discuss the validity of each of them.

Weighted linear regression

This analysis will be applied to the same transformed data that were used for the previous analysis. The idea of weighting is to prevent data points at the imprecise end of the graph having an excessive influence on the regression. The regression usually attempts to minimize the total squared residuals, but in a weighted regression the squared residuals are multiplied by weights where the weighting factors depend on the precision of the Y values. We still have to make assumptions about the system. In particular:

1. After transformation the variables are linearly related.

2. We can simulate equal, normally distributed precision in the transformed reaction rates by using different weighting factors at different pyruvate concentrations.

There are two strategies for deciding on weighting factors:

1. Mathematically. If heteroschedasticity is due to transformation of essentially homoschedastic data there may be a relationship between weighting and x which can be determined mathematically. This is true of our enzyme kinetics study. Without going deeply into the maths, the appropriate estimate of population standard deviation is σ is proportional to $1/v_0^2$.

2. Statistically. If you have replicate measurements you can determine a separate standard deviation for each selected value of x. Then s is used in the calculation of χ^2.

Take a look at the section above on residuals for a more detailed explanation of the calculations of residuals and weighting.

There is a serious doubt about statistics determined from weighted regressions. The problem is that the magnitude of the variation in Y can be manipulated with weighting factors, but if the original measurements were normally distributed, the transformed values will not be normally distributed. There may sometimes even be asymmetrically distributed values about the regression line (i.e. skewed). Weighting can fix the magnitude of spread either side of the regression line but

LIVERPOOL
JOHN MOORES UNIVERSITY
AVRIL ROBARTS LRC

can't change the shape of the spread. So there must be doubt associated with the coefficients determined by this method and if there is a type of regression possible that doesn't involve transforming normally distributed data, that is preferred. Popular spreadsheet software rarely has facilities for weighted or non-linear regressions and the following figures were obtained using Origin, a scientific statistical program produced by MicroCal. The inputs to the program are:

1. Selection of the equation of a straight line from the menu of equations.
2. The transformed data.
3. An initial guess at the slope and intercept.
4. A set of weights, one for each pyruvate concentration.

Three columns of figures were input into the software: 1/[pyruvate] (the independent variable), $1/v_0$ (the dependent variable) and $1/v_0^2$ (the relative standard deviation in v_0). The first guess at the slope was 10000 and the first guess intercept was 20.

With the first guess coefficients χ^2 was calculated;

At 0 iterations $\chi^2 = 2.6132 \times 10^{-6}$

After 1 iteration $\chi^2 = 4.3532 \times 10^{-8}$ (a fifty-fold improvement in fit)

After 2 iterations $\chi^2 = 4.3416 \times 10^{-8}$ (a slight improvement in fit)

Only two iterations of the software were needed because χ^2 couldn't be improved by further iteration. At this point the coefficients were presented by the software (with standard errors). Table 6.15 contains the coefficients that were found.

Table 6.15

	A	B	C
1		Coefficients	SE
2	Intercept	22.47099	0.0968
3	X Variable 1	7663.48083	57.39564

The slope is lower than it was for the non-weighted line fit and the intercept is a little higher. Why? A look at the residual plot, Figure 6.31, will help to explain.

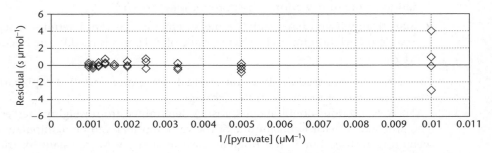

Figure 6.31 LDH reaction kinetics – double reciprocal residual plot. The points are slightly displaced compared with Figure 6.30.

The weighted regression has recognized that the points with the worst precision lie on the right and has placed greater importance (weight) on points to the left of the graph. That has dropped the regression line down on the right and lifted it up slightly on the y-axis. The V_{max} and K_m values are calculated from the coefficients in exactly the same way as for the non-weighted regression:

$$V_{max} = 0.044502 \ \mu mol \ s^{-1}$$

$$K_m = 341.038861 \ \mu M$$

Both these values are a little lower than for the non-weighted regression of the transformed data.

The software has provided us with coefficients of the straight line and confidence limits on those constants. It's easy to accept these statistical values and there is a dangerous tendency to trust them simply because a sophisticated computer program was used to calculate them. You have to remember that the software only implements the statistics you ask it to perform. The original untransformed dependent variable had error that was governed by the normal distribution. It is inevitable that after a reciprocal transform the error will no longer be normally distributed. If you are not convinced, Figure 6.32 may help.

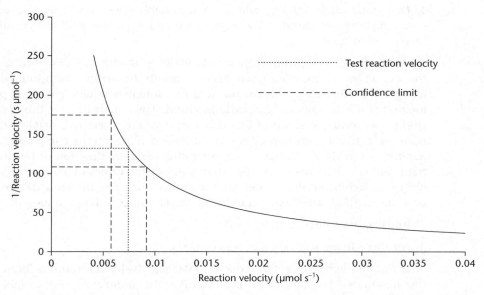

Figure 6.32 Effect of transformation on precision. This is a graph of the reciprocal of reaction velocity against reaction velocity. The dotted line shows a typical reaction velocity and how it maps onto a reciprocal. The dashed lines show the upper and lower 95% confidence limits on the reaction velocity and how they map onto reciprocals. If the precision in the determination of reaction velocity is normal, the confidence limits are equal distances either side of the measured value. You can see that the reciprocal transform gives rise to confidence limits that are not equal distances from the transformed measurement and you can conclude that imprecision is no longer normally distributed.

The confidence limits are equal distances either side of the reaction velocity, but the transformed confidence limits clearly show a skew to one side. The result of the skew is that negative errors in the dependent variable will have a bigger influence on the regression than they should.

Non-linear line fitting

Like every line-fitting method, assumptions are made about the data:

1. The variables are related according to the equation of a hyperbola. (A correct assumption with the range of substrate concentrations used in our experiment.)
2. Initial reaction rates are determined with equal, normally distributed impreci-sion. (Also a correct assumption for our enzyme.)

The input to the computer software is:

1. Selection of the equation of a hyperbola from the menu of equations.
2. The untransformed data.
3. The software is instructed not to use any weighting scheme. (It will make ex-actly the same calculations of χ^2 but with s set to 1.0 for all values of [pyruvate].)
4. First-guess values for V_{max} and K_m. A reasonable guess would set V_{max} equal to the highest rate found in the experiment and K_m to the median pyruvate con-centration used.

All least squares line-fitting software calculates χ^2 in the way described above in the section on residuals but programs can vary in the method used to modify the first-guess coefficients to obtain the second, estimated values. The most popular method is the Levenberg–Marquart algorithm. The details of how this algorithm works is beyond the scope of this text and if you're interested, detailed descrip-tions are available elsewhere. You can choose when to stop the computer program iterating – usually χ^2 reaches a minimum after no more than ten or twenty itera-tions and it is reasonable to stop when consecutive values of χ^2 are the same to five or six significant digits. Here are the results, analysing the same data as in the previous method, after four iterations of the Levenberg–Marquart alogrithm:

Initial guess values were $V_{max} = 0.0322$, $K_m = 500$

For these first-guess values χ^2 was 0.00009

After just four iterations χ^2 was reduced to 0.000000044 and was not reduced by fur-ther iterations. The values of the coefficients at that point are shown in Table 6.16.

Table 6.16

	A	B	C
1		Estimate	SE
2	V_{max}	0.04451	0.00019
3	K_m	341.4234	3.96319

Do the data fit the curve? Take a look at Figure 6.33.

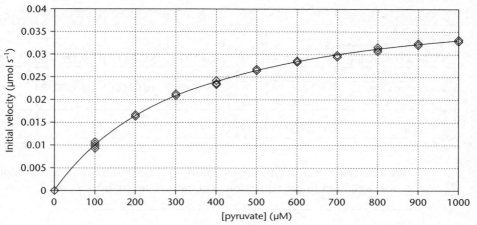

Figure 6.33 LDH reaction kinetics. It is difficult to judge scatter on this graph.

The residual plot, Figure 6.34, shows data evenly distributed above and below the line and there seems little doubt that the correct equation for the curve (i.e. a rectangular hyperbola) has been selected;

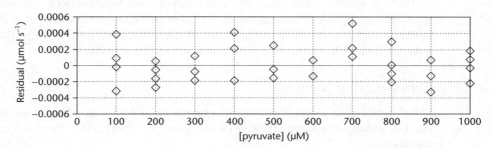

Figure 6.34 LDH reaction kinetics – residual plot. It is easier to see the variation between replicates in a residual plot.

Just as it was possible to test for linearity when the transformed data were fitted to a straight line, we can test statistically for a systematic deviation of the regressed line from a hyperbola. The only difference in the calculation is that y values must be predicted using the hyperbola and K_m and V_{max} instead of the equation of the slope and intercept of a straight line. Once the residuals are calculated the calculations are identical. The results of the analysis are in Table 6.17.

It is more than 99.9% likely than deviations from the given hyperbola are due only to imprecision in the determination of reaction velocities. This looks like very good news because it seems to confirm that the underlying system follows a hyperbola but you must remember the great danger of extrapolating from experimental data – perhaps if we extend the pyruvate concentrations higher than we did, the hyperbola model will break down.

Table 6.17

	A	B	C	D	E	F	G
1	ANOVA						
2	*Source of variation*	*SS*	*df*	*MS*	*F*	*P-value*	*F crit*
3	Between groups	0.002036	9	0.000226	5518.117	9.38E–46	2.210697
4	Regression	0.002036	1	0.002036			
5	Non-linearity	2.76E–08	9	3.07E–09	0.07482	**0.999838**	
6	Within groups	1.23E–06	30	4.1E–08			
7							
8	Total	0.002037	39				

We can also check for heteroschedasticity in the same way as was done with the non-weighted linear regression. The results are in Table 6.18.

Table 6.18

	A	B	C	D
1	Highest S^2	Lowest S^2	F_{max}	P
2	9E–08	1.33E–08	6.75	0.263509

There is a 26% chance that the difference in precision between the best and worst set of replicate measurements was due to random variation in a system where underlying precision is equal for all measurements. So we accept that the data are homoschedastic. And that is the end of the statistics!

Comparison of methods

You'll notice that the coefficients obtained from the various analyses are similar but not exactly the same. Table 6.19 compares the coefficients from the various methods.

Table 6.19

	Linear regression of transformed data	Weighted linear regression of transformed data	Non-linear regression of untransformed data
K_m	343.6	341.0	341.4
V_{max}	0.04464	0.04450	0.04451

How do you decide which is the correct answer? It is a principle of science that you must never select methods for analysis according to which provides results that fit your personal expectations. Instead, you should use your understanding of the analytical methods and use the method that is valid for the experimental situation in question. Taken one by one:

1. Linear regression of the untransformed data is rejected because the data are significantly non-linear.

2. Non-weighted, linear regression of transformed data is rejected because the transformed reaction rates are heteroschedastic and this breaks an assumption required for this analytical method.

3. Weighted, linear regression of transformed data is rejected because weighting may cause distribution of error that is not normal and this breaks an assumption required for least squares fitting.

4. Non-weighted, non-linear regression is accepted because all the assumptions required for this analytical method are valid for the experimental situation in question.

Table 6.20 summarizes the results from the one valid analysis.

Table 6.20

	A	B	C
1		Estimate	SE
2	V_{max}	0.04451 μmol/s	0.00019 μmol/s
3	K_m	341.4234 μM	3.96319 μM

Mathematics of non-weighted linear regression

It's possible that you may want to perform a non-linear regression at times when a computer isn't available. It isn't that difficult to use a pocket calculator to make the necessary calculations. Unlike non-linear regression, linear regression doesn't involve iterative computations. As with the calculation of standard deviation there is a formula that makes good sense of the calculation and another more obscure form of the formula that is easier to calculate. The slope of the line, m is:

$$m = \frac{\sum (X - \bar{X})(Y - \bar{Y})}{\sum (X - \bar{X})^2}$$

This would require you to find the means first, but the alternative formula is:

$$m = \frac{\sum XY - \dfrac{\sum X \sum Y}{N}}{\sum X^2 - \dfrac{(\sum X)^2}{N}}$$

The procedure is:

▶ Note down X and Y values in a column, making sure the correct Y value is paired with the correct X value.

▶ Add another column and calculate X^2 for each X value.

▶ Add another column and calculate the product XY for each value.

▶ Total up each column.

▶ Calculate m according to the formula above.

▶ Calculate $\bar{X} = \dfrac{\Sigma X}{N}$

▶ Calculate $\bar{Y} = \dfrac{\Sigma Y}{N}$

▶ Once you have the slope you can easily find the intercept c on the y-axis because the line that best fits the data always passes through a point defined by the two means, $\bar{X}$, $\bar{Y}$:

$$c = \bar{Y} - m\bar{X}$$

Summary

● The relationship between two variables can be classified as linear/non-linear, bivariate/univariate and homoschedastic/heteroschedastic. The choice of statistical method used to find parameters that describe the relationship is dependent on the type of relationship.

● Data are sometimes transformed before a line-fitting method is applied. Exercise 34 in Chapter 9 tests your ability to apply the appropriate transformation for various equations.

● In the absence of statistical software or pocket calculator, a graphical method can be used to fit a line to data. It is difficult to assess the reliability of coefficients determined this way. Exercise 32 asks you to plot a graph and fit a line by eye and Exercise 33 follows on to show you how you can assess the goodness of fit of your line.

● When a line has been fitted to data it is possible to use interpolation and extrapolation to predict one variable given the value of the other. Greater uncertainty is attached to extrapolated values than interpolated values. Exercise 36 asks you to fit a line graphically to data and both interpolate and extrapolate.

● There are assumptions made about data that apply for all type of least squares line fitting. There are additional assumptions for each type of least squares line fitting. Exercise 35 tests your ability to test the assumptions for one least squares method against different sets of data.

● Residuals are central to least squares line-fitting methods, and residual plots are also useful for the subjective assessment of the validity of the statistical method that was applied. The first part of Exercise 39 asks you to test your ability to judge subjectively the validity of the line fitting from a residual plot.

● Linear regression can be implemented with a pocket calculator but there is less likelihood of data entry errors if you use a computer. Example 10 in Chapter 8 shows how to use a spreadsheet program to implement a linear regression. Exercises 37 and

38 in Chapter 9 ask you to perform linear regression with a pocket calculator and with a computer respectively.

- A statistical test can determine the significance of non-linearity in a data set. Doing this with a spreadsheet is demonstrated in Chapter 8, Example 11, and you can test your ability to interpret the results of such an analysis with Exercise 39 in Chapter 9.

- A statistical test can determine the significance of heteroschedasticity in a data set. Doing this with a spreadsheet is demonstrated in Chapter 8, Example 12 and you can test your ability to interpret the results of such an analysis with Exercise 39 in Chapter 9.

- Weighting of data points can compensate for heteroschedasticity.

- Non-linear line fitting is usually an iterative process (the most popular algorithm is the Levenberg–Marquart method) and requires a computer or very sophisticated pocket calculator to implement. Ordinary spreadsheet software is usually inadequate for iterative line-fitting methods and specialist statistical software is needed. Example 13 in Chapter 8 shows how to use a typical statistical package on a computer to carry out non-linear regression.

Chapter 7

Screening for colorectal cancer

Introduction

The purpose of this chapter is to demonstrate the importance of probability for the communication of risk to medical staff and patients and its importance for decision making. The basic principle of combining mathematics and English effectively is applicable to all types of science, but it is possible that you have little experience of research science and so a straightforward medical case study has been selected. Before we get to the maths we need to look at the selected medical scenario that will be the subject of the case study.

Screening programmes

A medical screening programme is a two-stage programme that attempts to find people with a disease before they develop symptoms that would take them to a doctor of their own initiative. The first stage is to use an inexpensive, but perhaps partly unreliable, test on a mass population. Some people will get positive results but at this stage it is not certain that they all have the disease. In the second stage a reliable, but perhaps expensive test is used on everyone that got a positive result in the first stage. Screening programmes involve the participation of very large numbers of people who have no reason to suspect that they have any disease. The cost of the first stage test must be kept reasonably low because of the large numbers of people involved. Overall the programme remains effective because the second-stage testing is more reliable.

Colorectal cancer

Colorectal cancer is a diseased growth on the inside surface of the colon or the rectum. The tumour can develop quite slowly and may cause pain only at a late stage. Before there is any pain the tumour may bleed a little into the bowel. This disease is more common in affluent countries and it is generally considered that a diet lacking in fruit and vegetables is a contributing factor. This disease is more common in the elderly and is rarely seen in the young. Colorectal cancer is a candidate for a screening programme because it develops without initially causing symptoms and because treatment can be much more successful if started early. Thousands of people in the UK have colorectal cancer in its early stages but are completely unaware of it. When the tumour grows beyond a certain point these people will suffer from abdominal pain and will, it is to be hoped, visit their doctor. At some stage in the medical investigation the inside of the colon will be examined using an instrument called an endoscope. This is a flexible tube that is passed through the bowel and contains optics that allow a doctor to view the inside surface of the rectum and colon, a procedure called colonoscopy. The tumour will be visible and a diagnosis can be made, followed by appropriate treatment.

One way to reduce the number of deaths due to colorectal cancer would be to view the inside of the bowel of every adult at regular intervals. This is impractical for a number of reasons. It would be very expensive to implement and the discomfort and inconvenience of the procedure may mean that it is difficult to persuade people to participate, especially since the disease is quite rare. In fact there is some small danger from this procedure because it is possible to tear the bowel, and injury to the bowel can result in serious complications. The British Government (Department of Health) has decided (in 1998) to conduct large-scale studies to assess the feasibility of a national screening programme for colorectal cancer. This may result in a screening programme being implemented within a few years. If it is finally implemented it will be the first new national mass screening programme since breast cancer screening was introduced.

The faecal occult blood test

The faecal occult blood (FOB) test is a test for traces of blood in the faeces (occult means hidden). It is important to realize that this is not a direct test for colorectal cancer because there could be other reasons for blood to be found in the faeces: inflammation of the bowel lining due to aspirin use, haemorrhoids, contamination of the faecal sample with menstrual blood, and so on. There may also be substances in the diet which imitate blood chemically. On the other hand, some tumours will not bleed, or will bleed only intermittently, and there are substances in the diet that can interfere with the test to prevent a reaction with blood. These things will lead to negative results. Despite these drawbacks the test is simple to administer and cheap.

Screening with the faecal occult blood test

The question is, should the faecal occult blood test be used in a mass screening programme for colorectal cancer?

Mass application of the test is feasible, but only a proportion of positive results will be owing to colorectal cancer and some people with colorectal cancer will get negative results. In a screening programme, everyone who receives a positive result will have a follow-up colonoscopy, and so in the end the false positives from the faecal occult blood test will be distinguished from the true positives. There are four different outcomes of the screening programme for an individual:

1. True negative on the FOB test. A true negative is a negative result correctly given to a person without a tumour. This should be the biggest group.
2. False negative FOB test. Hopefully a very small group. These people will have no follow-up testing and no treatment but may go on to develop symptoms.
3. False positive FOB test. These will suffer the stress of a positive result for the FOB test but after an internal examination will be found to be perfectly healthy.
4. True positive FOB test. These will be diagnosed as having colorectal cancer and will get prompt treatment. Some of these people would have died if the disease had been allowed to progress to a point when it caused pain.

So what factors must be taken into consideration in deciding whether to implement a screening programme? Here is a list (in no special order):

1. The probability that a person who has a tumour will get a positive result from the FOB test.
2. The probability that a person who has a tumour will get a negative result from the FOB test.
3. The probability that a person who does not have a tumour will get a negative result.
4. The probability that a person who does not have a tumour will get a positive result.
5. The rate of incidence of tumours in the population.
6. The proportion of people tested who will get positive results and will need follow-up testing.
7. The proportion of people having follow-up testing who will be found to have colorectal cancer.
8. The proportion of people tested who will get negative results and will not be given follow-up testing.
9. The proportion of people with negative results who will actually have cancer.
10. The emotional toll on people of receiving positive results from the FOB test.
11. The reluctance of the public to suffer an internal examination when they don't feel ill.
12. The cost of the FOB test.
13. The cost of colonoscopy.
14. The number of lives that will be saved by the programme.

Some of these questions are interrelated and involve levels of uncertainty. Some of them can't be quantified and can only involve human judgement.

Modelling a screening programme using the FOB test

British studies into screening are not yet under way so data are not yet available that would enable us to judge the effectiveness of the FOB test in screening. However, we can make educated guesses about the reliability of the FOB test and about the rate of incidence of colorectal tumours. It's possible to pre-empt the real data and to model the system under study. Modelling means that you determine the mathematics that governs a system in general terms so that you can then ask 'what if' questions. For example, what if colorectal cancer were more common than it is now? Or, what if negative results from the FOB test were less reliable?

In the following text a type of study to determine the effectiveness of an FOB test is described. It isn't a perfect study and you will be able to see faults in it, but it provides an opportunity to learn some important principles of data handling and decision making based on data. Since this is a model of a study that hasn't yet been done in the UK the figures are made up. The figures that govern the reliability of the FOB test are a rough average of results from real studies into a variety of different FOB tests. (Faecal occult blood test kits from different manufacturers have different levels of reliability.) More doubt exists around the rate of incidence of colorectal tumours in the UK population. In the model presented here, we take as a given that *the probability of individual taken from the population at random will have colorectal cancer is 0.001 or 0.1%*. This is a very rough estimate.

It's important that you treat the whole of this exercise as a 'what if' exercise. To get definite answers to base policy decisions on in the UK we would need real data.

What if the FOB test gives a positive result to 99% of people with tumours, gives a positive result to 98% of people without tumours and colorectal tumours are quite rare.... ?

Experimental design

Imagine we have developed a new method for doing an FOB test, the CraPhaem test, and we want to find out how effective it is. We have a budget that limits the number of volunteers we can have in the study to 2000 people, but since the disease is quite rare, if we select these 2000 randomly from the general population we might not get any volunteers with colorectal cancer. That's a shame because the mathematics would be much simpler if we could just take 2000 people and treat them as a microcosm of the whole population. The alternative is to arrange for separate testing of a group of 1000 people who have colorectal tumours and of 1000 people who don't have tumours. That way it should be possible to get an estimate of the reliability of the test for these two different groups and with the appropriate maths we can combine the two sets of figures to answer other questions about a screening programme.

We can confirm that each of the volunteers in the study has or hasn't got the disease by viewing their colon with the endoscope, and we will also apply the FOB test and obtain a positive or a negative result for each. When we go on to analyse the results it will be important that the two groups of figures, those from people with tumours and those from people without tumours, are segregated. This segregation is necessary because in the study half the people have tumours and this is not representative of the tiny number of people in the country who have tumours. Table 7.1 shows the data from the study.

Table 7.1		
	Positive	Negative
With tumours	985	15
No tumour	19	981

These data tell us nothing about how many people in the population have colorectal tumours so we would have to rely on figures from someone else's study. We will take the proportion of the population with colorectal tumours as 0.001. We now have all the data we need to start making calculations; but before we start we need to choose some symbols that will allow us to make concise notes.

Notation

We're going to be looking at a number of different probabilities, and if we're not careful we could end up with a lot of confusion. Therefore, to avoid long and unwieldy statements of risk while we work with the figures we should adopt some mathematical symbols. This will allow us to combine probabilities and note down these calculations concisely. Our most important symbols will be A and B.

A = The statement: Patient has a colorectal tumour.

B = The statement: Patient has a positive result on a faecal occult blood test.

Since we are interested in probabilities we will use P( ) in combination with the other symbols to indicate a probability. Table 7.2 contains a complete list of all combinations of these symbols along with an English translation for each.

Each of these sixteen probability figures can be calculated from either the original data or each other. The arithmetic you need is quite simple.

Initial calculations

People with tumours

Our first calculation will look at the effectiveness of the FOB test when applied to people that we know have tumours. We just want to know the proportion who will get positive results and who will get negative results. We take the data from volunteer cancer sufferers as being representative of all cancer sufferers. (That may be a dubious assumption but we will continue.)

Table 7.2

Symbol	Statement in English
P(A)	The probability that an individual has a colorectal cancer (when status as regards a faecal occult blood test is not known)
P(B)	The probability that a person has a positive result for a faecal occult blood test (when status as regards colorectal cancer is unknown)
P(NOT A)	Probability that a patient does **not** have a colorectal tumour (when status as regards a faecal occult blood test is not known)
P(NOT B)	Probability that a patient has a **negative** result on a faecal occult blood test (when status as regards colorectal cancer is unknown)
P(A|B)	Probability that a patient has a colorectal tumour (**given** that we know that the patient has a **positive** result from a faecal occult blood test)
P(A|NOT B)	Probability that a patient has a colorectal tumour (**given** that we know that the patient has a **negative** result from a faecal occult blood test)
P(NOT A|B)	Probability that a patient does **not** have a colorectal tumour (**given** that we know that the patient has a **positive** result from a faecal occult blood test)
P(NOT A|NOT B)	Probability that a patient does **not** have a colorectal tumour (**given** that we know that the patient has a **negative** result from a faecal occult blood test)
P(B|A)	Probability that a patient will have a **positive** result from a faecal occult blood test (**given** that we know they have a colorectal cancer)
P(B|NOT A)	Probability that a patient will have a **positive** result from a faecal occult blood test (**given** that we know they do **not** have a colorectal cancer)
P(NOT B|A)	Probability that a patient will have a **negative** result from a faecal occult blood test (**given** that we know they have a colorectal cancer)
P(NOT B|NOT A)	Probability that a patient will have a **positive** result from a faecal occult blood test (**given** that we know they do **not** have a colorectal cancer)
P(A AND B)	Probability that a patient has a colorectal cancer **and** a **positive** result on a faecal occult blood test
P(A AND NOT B)	Probability that a patient has a colorectal cancer **and** a **negative** result on a faecal occult blood test
P(NOT A AND B)	Probability that a patient does **not** have a colorectal cancer **and** has a **positive** result on a faecal occult blood test
P(NOT A AND NOT B)	Probability that a patient does **not** have a colorectal tumour **and** has a **negative** result on a faecal occult blood test

$$P(B|A) = \frac{\text{Cancer sufferers with positive results}}{\text{Total cancer sufferers}}$$

$$= \frac{985}{1000}$$

$$= 0.985$$

$$P(\text{NOT } B|A) = \frac{\text{Cancer sufferers with negative results}}{\text{Total cancer sufferers}}$$

$$= \frac{15}{1000}$$

$$= 0.015$$

Alternatively:

$$P(\text{NOT } B|A) = 1 - P(B|A)$$

$$= 1 - 0.985$$

$$= 0.015$$

This is not particularly useful information for people who may take part in a screening programme but is useful for those who may run a screening programme.

People with no tumour

We can make the same basic set of calculations but this time using data from the group of volunteers without a tumour.

$$P(B|\text{NOT } A) = \frac{\text{Healthy people with positive results}}{\text{Total healthy people}}$$

$$= \frac{19}{1000}$$

$$= 0.019$$

$$P(\text{NOT } B|\text{NOT } A) = \frac{\text{Healthy people with negative results}}{\text{Total healthy people}}$$

$$= \frac{981}{1000}$$

$$= 0.981$$

Alternatively:

$$P(\text{NOT } B|\text{NOT } A) = 1 - P(B|\text{NOT } A)$$

$$= 1 - 0.019$$

$$= 0.981$$

We now have four of the sixteen probabilities so we're on our way to cracking this problem.

The population at large

The previous sets of figures tell us little about how the test will perform in a screening programme because they relate to people whom we know have cancer and to people we know don't have cancer. In a screening programme we don't know which people have tumours and we don't know in advance what test results people will get. We need to calculate four probabilities – one for each of the four outcomes an individual might fall into (as discussed earlier). Let's take P(A AND B) first.

This is a proportion of a proportion because first we look at the proportion of the population that have tumours and then within that group the proportion that will get positive results. So in symbols:

P(A AND B) = P(A) × P(B|A)

P(A) is the given, 0.001, so the calculated figure must be small. Very few people in the population have tumours and not all of them will get positive results:

P(A AND B) = 0.001 × 0.985 = 0.000985

So it's just a matter of taking the right probability calculated from the study data and scaling it to the rate of incidence of the disease. The other three probabilities in this section are calculated in a similar way, so let's construct a table (Table 7.3).

Table 7.3

	A	NOT A
B	P(A AND B)	P(NOT A AND B)
NOT B	P(A AND NOT B)	P(NOT A AND NOT B)

Table 7.4 is filled in with the appropriate formulae. Table 7.5 contains the actual calculations, and finally Table 7.6 has the answers.

Table 7.4

	A	NOT A
B	P(A) × P(B|A)	P(NOT A) × P(B|NOT A)
NOT B	P(A) × P(NOT B|A)	P(NOT A) × P(NOT B|NOT A)

We've reached the half-way stage with the calculations and it is important that you don't go any further until you have fully understood things so far. First, you should review the calculations that have been made and make sure you understand them. Second, go back again and look at the sizes of the numbers and relate them back to the screening programme. You've got three different situations to look at: probabilities that predict what will happen when people with tumours

Table 7.5		
	A	**NOT A**
B	0.985×0.001	0.019×0.999
NOT B	0.015×0.001	0.981×0.999

Table 7.6		
	A	**NOT A**
B	0.000985	0.018981
NOT B	0.000015	0.980019

have the test, probabilities that predict what will happen when people without a tumour have the test, and then probabilities that predict what will happen when people who may or may not have tumours have the test.

Totals

Although these last four probabilities give you the four groups in a screening programme, when you actually run the screening programme you won't find out immediately which people have the tumour, just who has positive results and who has negative results. It will be useful to know the total number of people who will have positive results and the total with negative results. We can just expand on the table from the previous section (Table 7.7).

Table 7.7			
	A	**NOT A**	**Total**
B	P(A AND B)	P(NOT A AND B)	P(B)
NOT B	P(A AND NOT B)	P(NOT A AND NOT B)	P(NOT B)
Total	P(A)	P(NOT A)	1.0

Just to explain one of these figures in symbols:

P(B) = P((A AND B) OR (NOT A AND B))

= P(A AND B) + P(NOT A AND B)

The answers are in Table 7.8.

Table 7.8			
	A	NOT A	Total
B	0.000985	0.0108981	0.019966
NOT B	0.000015	0.980019	0.980034
Total	0.001	0.999	1.0

People with positive results

The previous sets of figures predict what will happen to an individual entering the screening programme when you know nothing about them. But how should you advise someone who now has a positive result and what action should you take? None of the probabilities you have so far helps with that. So, in symbols what is the probability that you need to calculate? It is P(A|B), which in English is the probability that an individual has a tumour *given that* their test result is known to be positive. Alternatively, we may also want to know P(NOT A|B), which is the probability that an individual does *not* have a tumour *given that* their test result is known to be positive. The important point here is that we want to look only within the group of people that will get positive results, so we are focusing on the top row of Table 7.8. Let's reason through the symbols carefully.

We know already that:

P(A AND B) + P(NOT A AND B) = P(B)

And since a person with a positive result can only fall into two groups, with or without a tumour, we can also conclude that:

P(A|B) + P(NOT A|B) = 1.0

In fact the two probabilities we want are just scaled up from the corresponding probabilities in Table 7.8. So:

$$P(A|B) = \frac{P(A \text{ AND } B)}{P(B)}$$

and

$$P(NOT\ A|B) = \frac{P(NOT\ A \text{ AND } B)}{P(B)}$$

Putting the figures into these calculations we get:

$$P(A|B) = \frac{0.000985}{0.019966}$$

$$= 0.049333868$$

$$P(\text{NOT } A \mid B) = \frac{0.018981}{0.019966}$$

$$= 0.950666132$$

We have a surprising result here. When someone gets a positive result in this screening programme there is only a 5% chance that they actually have cancer. Put another way, most people (95%) with a positive result will *not* have a tumour. The right calculations have been made, this is definitely the right answer, but why is this last figure so high when at the first stage of these calculations the test looked quite reliable? The key to this question is knowing that most people in the screening programme do not have a tumour. Let's reason it through step by step and as you read this refer back to the figures that support each of these statements:

1. A small proportion of healthy people will get a positive result (false positive).
2. Most who are tested will not have a tumour.
3. In other words, few people in the population have a tumour.
4. Since the disease is quite rare, no matter how effective the test the number of genuine positive results in the screening programme will be very small.
5. A small number of people with false positive results is still a much bigger number than that very small number of people with true positive results.

*This type of result is called the **Paradox of the False Positive**. It helps us to make a very important point about handling data. You must avoid trying to make decisions by just looking at the raw data. Instead, you should ask the right questions and make sure you calculate figures that provide answers to those questions. If you hadn't gone through this maths you may have assumed that most people getting positive results would have cancer and you might have planned for many more people needing treatment than was correct, and if you didn't communicate the unreliability of positive results to people having the test, you would have caused extreme distress for thousands of people with false positives.*

People with negative results

We now have to think about people with negative results on this test. It's the same type of calculation as for positive results but we're looking at the second row in the table of figures for the whole population (Table 7.8 above).

$$P(A \mid \text{NOT } B) = \frac{P(A \text{ AND NOT } B)}{P(\text{NOT } B)}$$

and

$$P(\text{NOT } A \mid \text{NOT } B) = \frac{P(\text{NOT } A \text{ AND NOT } B)}{P(\text{NOT } B)}$$

Putting the figures into these calculations we get:

$$P(A|NOT\ B) = \frac{0.000015}{0.980034}$$

$$= 0.0000153056$$

$$P(NOT\ A|NOT\ B) = \frac{0.980019}{0.980034}$$

$$= 0.999984694$$

Put into English, the probability that someone has cancer given a negative result is 0.0015%. The chance that they don't have cancer, given a negative result is 99.9985%. This is a much more satisfactory result. Virtually all negative results will be true negatives. This is only the case because most people going into the screening programme will be perfectly healthy.

Complete modelling of the screening programme

You will have to wait for the results of the British screening programme before you know the real answers to the various probabilities calculated here from 'educated guess' data. All of the calculations above worked from a single set of data for the modelled study. What if colorectal tumours are more common than our estimate? What if colorectal tumours are less common than our estimate?

You can extend the modelling of the screening programme simply by using different input figures and running through exactly the same calculations. The input variables of the model are $P(A)$, $P(B|A)$ and $P(B|NOT\ A)$ and all the other probabilities can be calculated from them. A spreadsheet program is ideal for the purposes for modelling because you can create formulae that will update if you edit any of the cells containing variables. You might like to try this yourself. You can try all permutations of the following possibilities:

1. The test is good/poor at picking up people with tumours. $P(B|A)$ is high/low.

2. The test is good/poor at picking up negatives. $P(B|NOT\ A)$ is high/low.

3. The disease is common/rare in the population. $P(A)$ is high/low.

If you model all possible outcomes of the study and whatever the outcome, you would not recommend a national mass screening programme then there is not much point spending money doing the real study. Models are also valuable for answering 'what if' types of questions. You might want to run a model after a screening programme is implemented to predict what benefits will come if an improved test were used in the programme for example. Models are also valuable educational tools and if you can spare the time to model this case study with a spreadsheet you might find the Paradox of the False Positive is easier to believe in. See Example 3 in Chapter 8.

Graphical summary

It might help your understanding to model the problem graphically using simplified figures. Each of the graphs in Figures 7.1 to 7.8 divides a unit square into areas proportional to the appropriate probabilities. For probabilities concerned with having cancer the square is divided vertically, and for probabilities concerned with a test result the square is divided horizontally. In the one situation where both are involved the square is divided horizontally and vertically into four areas. The beauty of these graphs is that you can treat all the calculations we have done so far as simple geometry problems.

For the purposes of these diagrams the three probabilities that determine all the others have been set to values that are easier to work with (they will still be illustrative of the Paradox of the False Positive):

P(A) = 10%

P(B|A) = 20%

P(B|NOT A) = 80%

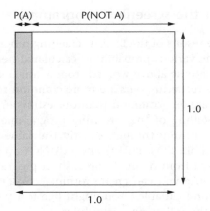

Figure 7.1 Incidence of tumours. This graph divides the population between those who have tumours and those who do not.

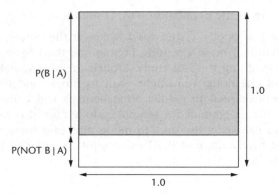

Figure 7.2 Test results for people with tumours. This graph is obtained from study data. The raw data are simply scaled to fit a total of 1.0. This is because we are interested in the proportion of results in each group.

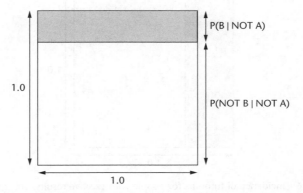

Figure 7.3 Test results for people with no tumour. This graph is obtained from study data using the same sort of mathematics as in Figure 7.2 but with different data.

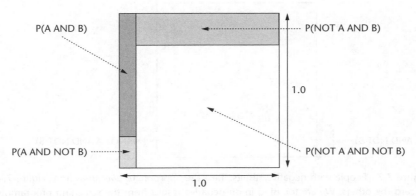

Figure 7.4 Incidence of tumours and test results within a screening programme. This graph is obtained by simple geometry. Figures 7.2 and 7.3 are narrowed to fit within the two areas of Figure 7.1. In other words, they are scaled to match the incidence of tumours, and consequently the column for people with tumours is narrow and the column for people without a tumour is wide.

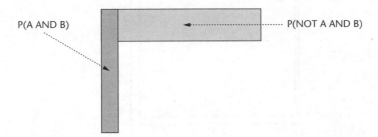

Figure 7.5 People with positive results. Take the appropriate two areas from Figure 7.4 and discard the others. We are focusing in on the positive results from the screening programme. The total area is equal to P(B).

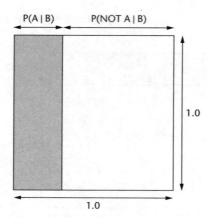

P(A | B)   P(NOT A | B)

1.0

1.0

Figure 7.6 Incidence of tumours for people with positive results. This graph takes the two areas in Figure 7.5 and scales them up to fit within a unit square. So this graph shows you how the group of people who have a positive result are divided into those who have tumours and those who do not have a tumour.

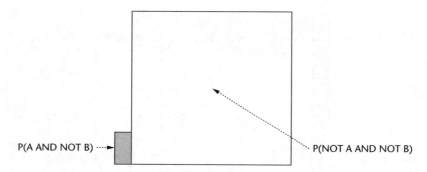

P(A AND NOT B) ⋯▶   ⋯▶ P(NOT A AND NOT B)

Figure 7.7 People with negative results. Take the appropriate two areas from Figure 7.4 and discard the others. We are focusing in on negative results from the screening programme. The total area is P(NOT B).

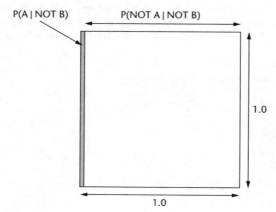

P(A | NOT B)   P(NOT A | NOT B)

1.0

1.0

Figure 7.8 Incidence of tumours for people with negative results. This graph takes the two areas in Figure 7.7 and scales them up to fit within a unit square. So this graph shows you how the group of people who have a negative result are divided into those who have tumours and those who do not have a tumour.

If you have the type of mind that works better with pictures than symbols you may find that this provides a much better explanation of the results. Go back over the diagrams, see how each follows on from the previous and then relate each to the screening programme itself. (Remember, though, that different figures have been put in to make sure that the areas are all big enough to visualize.) You can summarize the effect of the screening programme on a typical participant by picking out the three diagrams that assess the probability of having a tumour.

This series of questions in Figure 7.9, with graphical solutions, illustrate the ultimate value of the FOB test, i.e. the effect it has on your answer to a person's question, 'How likely is it that I have a colorectal tumour?' Having the test will show either an increased or a decreased risk that this person has a colorectal tumour.

Making decisions about a screening programme

Assume for the moment that the set of figures we calculated is a good prediction of the outcome of the real studies. You have some numbers, but that can't be the end of the story. You have to decide what the important issues are for implementing a screening programme and find the figures that address them. Here are some observations, although the list is far from complete:

1. The disease is fairly rare and so no matter how good the screening programme is there is a limited opportunity to save lives.

2. Quite a high proportion of people will get positive results and these must all have follow-up examinations. This is going to be very expensive.

3. The emotional trauma of getting a positive result and then the small but real danger of a colonoscopy will be inflicted on very many perfectly healthy people.

4. It might also be difficult to persuade people to have an internal examination when you know in advance that most will not have cancer. (Unless you deliberately exaggerate the risk.)

5. There will be a very small number of people receiving false negative results. This is a crucial feature of a good screening programme.

And now some observations about what we still don't know:

1. We only have figures that apply when we test the general population.

2. We know that incidence of tumours is higher in the elderly but we don't have a set of probabilities that apply specifically to the elderly.

3. Elderly people may also have thinner bowel linings and be more at risk of injury during a colonoscopy.

4. We don't have data for people who have symptoms of colorectal cancer. Is there any point using the test on people with abdominal pain and constipation or diarrhoea? We might mislead people who get negative results if we ignore their symptoms and tell them the risk of having cancer is very low.

Question	Answer
'I've not had any test and I don't have any symptoms. How likely is it that I have a colorectal tumour?' The answer is in the shaded area of the graph	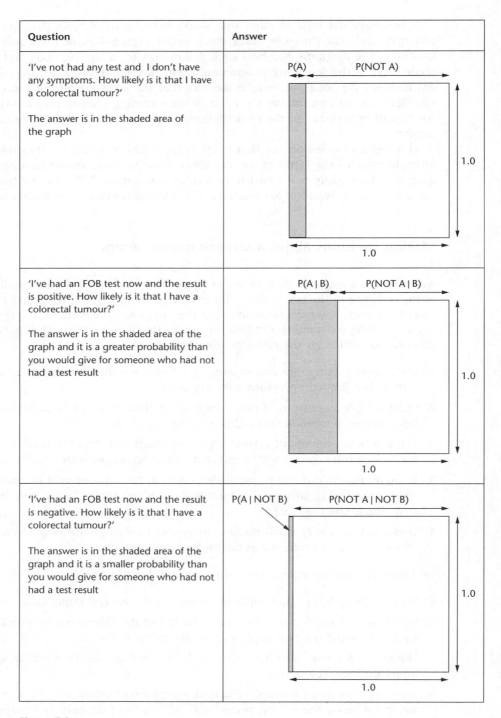
'I've had an FOB test now and the result is positive. How likely is it that I have a colorectal tumour?' The answer is in the shaded area of the graph and it is a greater probability than you would give for someone who had not had a test result	
'I've had an FOB test now and the result is negative. How likely is it that I have a colorectal tumour?' The answer is in the shaded area of the graph and it is a smaller probability than you would give for someone who had not had a test result	

Figure 7.9

In the UK we don't have a screening programme for colorectal cancer and this is probably due to two factors: the high cost and the judgement that few people would be prepared to suffer an internal examination if they have no symptoms, even if they had a positive result from a faecal occult blood test. The studies now sponsored by the Department of Health may lead to a reassessment. In the USA things are different for two reasons. First, screening programmes are less to do with central planning because there is no national health service. This means that colorectal cancer screening is a matter for individual choice and the government, in the form of the National Institutes of Health, can only advise the public to request testing. The second difference is that the individuals must pay for their own testing. So in the USA people are advised to have a colonoscopy annually after the age of 60, but only the relatively wealthy are likely to participate because this examination may cost them as much a $1000 per year. Commercial companies that manufacture FOB tests are likely to promote the use of their product with the intention of maximizing revenue. This is a very worrying situation because the motivation of the organization doing the testing is profit, and patient welfare is only of interest where it impacts on profit. Consider this moral dilemma. You are a scientific adviser to a life insurance company that is devising a policy aimed at elderly people. Would you insist that applicants have an FOB test or a colonoscopy before calculating a premium for them? What would you do if an applicant suffered a perforated colon as part of the colonoscopy? Pay them compensation or reject their application?

Screening for cervical cancer

In a way it is surprising that the UK National Health Service hasn't considered a colorectal cancer screening programme earlier, because in many ways the issues are similar to the cervical cancer screening programme that it has been running for many years now.

Similarities between colorectal and cervical cancer screening

1. Cervical cancer is also a rare disease. (Actually even more rare than colorectal cancer.)
2. Most 'positive' results from a smear test are false positives. (These reasons are discussed below.)
3. The test is unpleasant and invasive and it could be difficult to get the public to participate.

It is likely, although not proved, that cervical cancer starts with an infection by the virus that is also responsible for genital warts. Some of the cells on the surface of the cervix become abnormal – this is known as dyskaryosis. Usually a woman's natural immunity destroys these abnormal cells and things return to normal, but it is possible that cells showing dyskaryosis may multiply and develop into cancer. These cancers tend to develop quite slowly. It is extremely rare for a smear test to reveal cancer because most women have regular smear tests and are treated before cancer ever develops. So a positive test result simply means that cells showing dyskaryosis have been found. Cervical cancer is always preceded by cells with dyskaryosis *but* dyskaryosis does *not* always develop into cancer. This is analogous to the faecal occult blood test – the test used in the first stage of screening does not directly indicate cancer but indicates a higher risk.

Differences between colorectal and cervical cancer screening

1. With cervical cancer screening the positive result indicates an increased risk that cancer *will* develop in the future whereas with colorectal cancer screening the positive result indicates an increased risk that the patient *already* has colorectal cancer.

2. A positive FOB test will lead to colonoscopy which, although invasive and uncomfortable, doesn't involve any physical trauma to the bowel. If women have a positive result on a smear test they usually will have their cervix examined in a procedure called colposcopy. Unfortunately there is rarely much to learn by just looking at the cervix and colposcopy often ends with a biopsy. In a biopsy some of the cervix is cut away surgically and this can be painful and there is risk of infection of the wound. The tissue taken in the biopsy is examined in essentially the same way that the smear was. If there are significant amounts of abnormal cells in the biopsy, treatment will be recommended.

3. The test is very expensive because it involves a health worker's time to collect the sample and an expert technician to spend time preparing and examining each sample.

It's important to realize that women with some cervical cells showing dyskaryosis are not ill and this means that most women who get positive results on smear tests are perfectly well. Women with positive smear tests will usually be subjected to biopsies and possibly treatment, but in most cases the problem with dyskaryosis of cells would have cleared up if they had just been left alone. On the other hand, negative results are extremely reliable. If the woman actually has cancer it is almost certain that the smear will indicate this: if the woman has many cells showing dyskaryosis this is very likely to be discovered. If the woman has only a few cells showing dyskaryosis it is possible that this might be missed *but* there is a fairly low chance of this turning into cancer and if it does develop further it will happen slowly and be picked up on the next smear. So a negative result from a smear test is very reassuring, but a positive result does *not* mean the patient has cancer – it means there is a risk that she will develop cancer in the subsequent months.

What complicates assessing the value of the screening programme is that the disease seems to be very rare partly because one doesn't know when cancers have been prevented by giving treatment. You don't know just how many women would have gone on to develop cervical cancer if no action had been taken over their positive result on a smear.

How has the NHS managed to get so many women to participate in the cervical screening programme? The screening programme is probably effective but it hasn't been presented to women on its true merits because it would be difficult to explain risk assessments with all the maths associated with it. Instead, individual doctors have exploited people's fear of cancer by exaggerating the risk of getting cervical cancer. Another tactic of doctors to encourage participation is to get women used to smear tests at an early age. It's extremely rare for 16-year-old girls to get cervical cancer (virtually impossible if they are not yet sexually active) but most doctors will start doing smear tests when a teenager first asks for contraceptive advice. Young women are less likely to argue with the demands of their doctor, and once they get started on smear tests they are likely to continue. Another way that women are persuaded to comply is the language used with the smear test results. Women with positive results are told they have pre-cancerous cells (remember this just means cells with dyskaryosis have been found). Many women have pre-cancerous cells in their cervix and the term is only used to indicate the type of cell that in a very small number of women *may* become cancerous over a number of months. However, the term gives women the impression that they actually have cancer in its early stages and have been cured. Many women will have friends that have had positive results from a smear test and so they have a false impression that cervical cancer is more common than it really is.

This may all seem a diversion from the data handling that is the topic of this chapter but you always need to remember that the data and the calculations are just tools to help you make decisions and there are other issues to think about:

1. What is the best policy to maximize the welfare of the public?
2. Thousands of women every year will get positive results and have biopsies and possibly treatment when they would have been perfectly healthy and happy without intervention. To what extent is the emotional trauma for hundreds of thousands of women balanced against the hundreds of lives saved by the programme?
3. Could the emotional trauma be minimized without cancelling the programme?
4. What is the best way to persuade women to participate in a cervical screening programme?

There is one final moral to this story. It is very important that health policy is based on valid data and proper analysis of that data but clinical staff have a tendency to become so focused on things that can be quantified they can find themselves ignoring factors that can't be quantified. In the cervical screening programme it has been possible to assess risks, check the reliability of the screening laboratories, look at survival rates etc. and this has been the focus of medical professionals. What has been ignored by many in the past is the unquantifiable emotional trauma suffered by the thousands of women who are given positive results every year. This attitude is slowly changing and some health workers are reviewing the way they run screening programmes to address the welfare of healthy people as well as those with diseases.

Chapter 8

Data handling with a computer

Introduction

All the other chapters have worked on the assumption that most of the data handling you will do in your professional life will be executed using a computer. It is possible though that you don't yet have the experience with the type of computer software that is used for processing data. So the aim of this chapter is to present worked examples of computer-based data processing. All but one of the worked examples uses instructions specific to Microsoft Excel '97. Excel is a spreadsheet program and can be considered a multipurpose number processor. Excel has a number of statistical analyses built in but is weak on facilities for presenting and processing *x–y* data, so the last worked example uses Origin 5.0. Origin is a program that is designed specifically to work with scientific data, and among other things, is capable of non-linear line fitting, with or without weighting, using the Levenberg–Marquart algorithm. It is assumed that if you intend to try these worked examples on your own computer that you will first study the basic instructions on entering data and formulae that come with the software. (You can type a question such as 'How do I enter data into a cell?' into Office Assistant.)

Example 1 Simple data processing

A great strength of spreadsheets is that a single calculation in a cell that processes a single datum can be duplicated across a row or down a column almost instantly and you can therefore process a lot of data easily. In the following worked example, raw data from an LDH reaction kinetics run like the one presented in Chapter 6 are converted into reaction velocities. The experiment measures the rate of consumption of NADH at various concentrations of pyruvate. So just as in an assay for LDH, the raw data consist of a starting and ending absorbance but the starting absorbance will be the same for every run.

▶ Start with a new, empty workbook file. You get a new one when Excel starts up but you can also click on the New Workbook button on the toolbar.

Example 1   Simple data processing   163

▶ Enter the starting absorbance, the molar absorbtion coefficient of NADH, the volume of the reaction mixture and the duration of the reaction run, with labels at the top of your spreadsheet to match Figure 8.1.

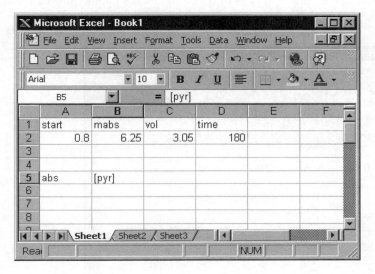

Figure 8.1

Sometimes it is useful to name a cell. The benefits will become clear later.

▶ Select cell A2 and then select the Name command on the Insert menu. From the submenu choose the command Define....

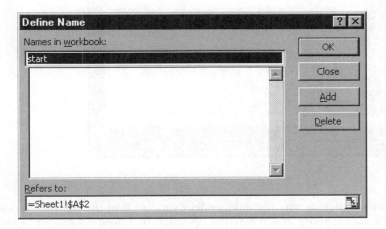

Figure 8.2

▶ The dialogue box, shown in Figure 8.2, cleverly guesses that the name you want to use is 'start' so all you have to do is click on the OK button.

▶ Do the same thing to name the other three cells 'mabs', 'vol' and 'time'.

Now you have to enter the raw data. This is not difficult.

▶ Select cells one by one and type in data and labels so your spreadsheet matches Figure 8.3.

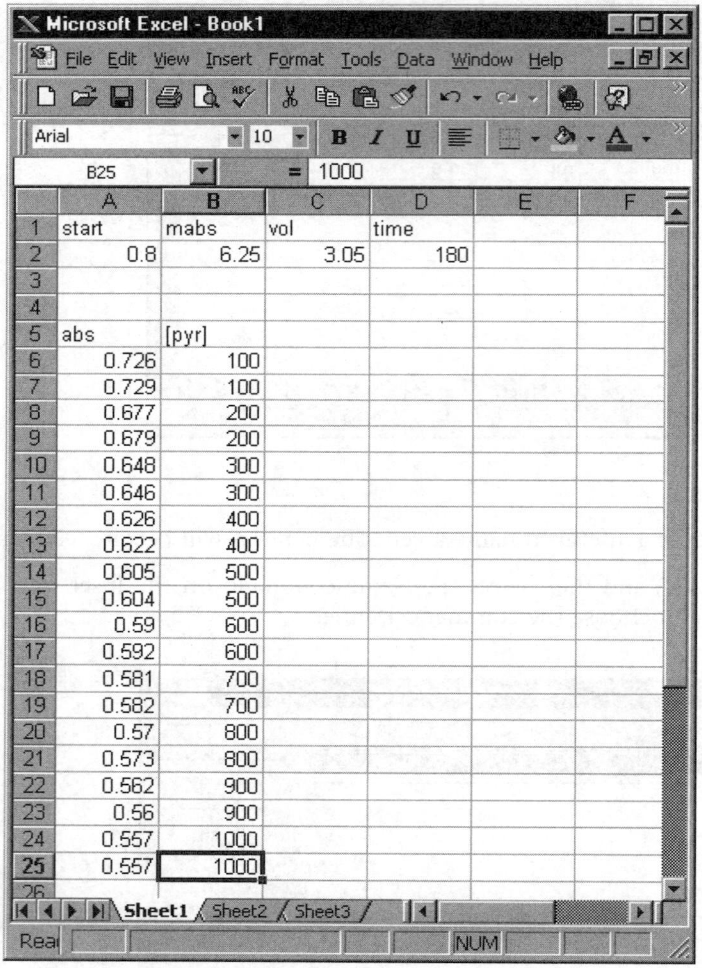

	A	B	C	D	E	F
1	start	mabs	vol	time		
2	0.8	6.25	3.05	180		
3						
4						
5	abs	[pyr]				
6	0.726	100				
7	0.729	100				
8	0.677	200				
9	0.679	200				
10	0.648	300				
11	0.646	300				
12	0.626	400				
13	0.622	400				
14	0.605	500				
15	0.604	500				
16	0.59	600				
17	0.592	600				
18	0.581	700				
19	0.582	700				
20	0.57	800				
21	0.573	800				
22	0.562	900				
23	0.56	900				
24	0.557	1000				
25	0.557	1000				

Figure 8.3

You may be wondering why the independent variable is in the right-hand column. It's because we are going to add reaction velocities in a third column. It's now time to enter a calculation for the first reaction velocity (Figure 8.4).

▶ Select cell C6 and enter:

=(start-A6)*vol/(mabs*time)

Example 1   Simple data processing   165

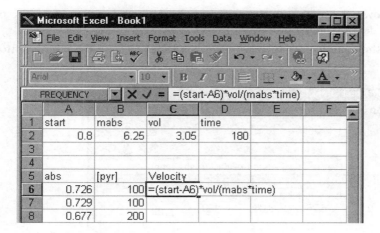

Figure 8.4

▶ Type the Return key to complete the formula and you should see the result of the calculation (Figure 8.5).

Figure 8.5

Now you could enter similar formulae into the other cells of the column but there is a quicker and simpler way to achieve the same effect. Look at the black selection box around cell C6. There is a small black square in the bottom right-hand corner. This is called the *fill handle*.

▶ Move your mouse pointer over the fill handle. You will see that the pointer changes to a small cross. This takes a steady hand and plenty of room on the table in which to move the mouse.

▶ Press down the mouse button and drag the fill handle down. Keep the mouse button down while you drag.

▶ When the column, down to cell C45, is selected, release the mouse button.

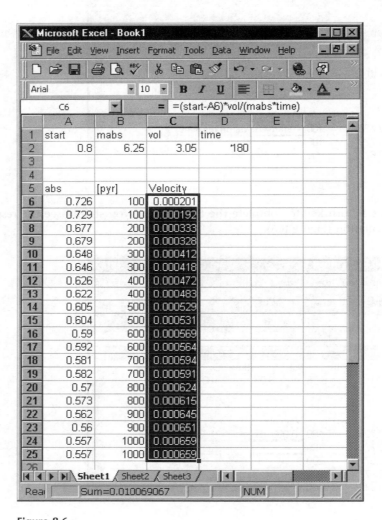

Figure 8.6

The whole column is filled with results, as shown in Figure 8.6. You can see what Excel has done by clicking on different cells in the column. The formula of the selected cell will be shown in the formula bar. Excel has filled the column with the same formula you entered into the first cell but it has altered each one so that it refers to the correct absorbance value. However, because you gave the constants special names they were *not* altered for each cell.

Example 1 Simple data processing   **167**

The ordinary A1 type of cell *reference* is called *relative* because it is relative to the cell where the formula is found. (In your spreadsheet the formula refers to the cell two cells to the left.) The *names* refer to the same cell or cells so they are called *absolute*.

▶ Save your file onto a suitable disk drive before you go any further. Computers sometimes crash and you lose all your work back to when you last saved the file.

Correcting mistakes

Imagine you had mistyped a datum. The absorbance at cell reference A15 should be 0.606. If you were using a calculator you would have to redo the concentration calculation. (The chances are that you would never have noticed the mistake on the calculator – that's another big advantage of using a spreadsheet.)

▶ Select cell A15 and type:

0.606

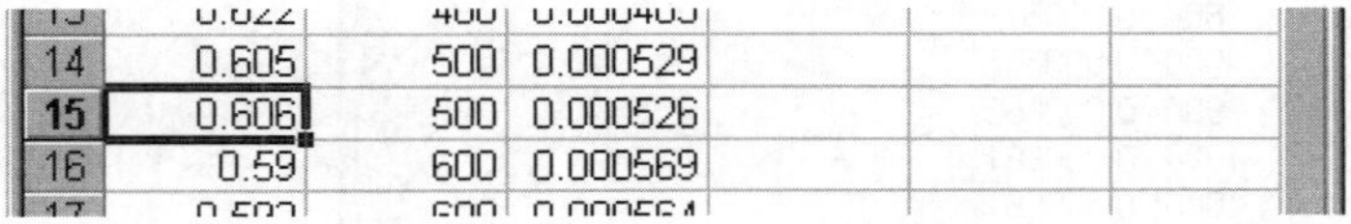

Figure 8.7

The correct concentration is instantly displayed (see Figure 8.7). There is a key feature of spreadsheets.

When you make an alteration in a spreadsheet all of the cells are checked and if necessary recalculated.

What if you had made a more serious error? If you used the wrong molar absorbance on your calculator you would have to redo *every* concentration calculation.

▶ Select cell B2 and enter:

6.23

Every reaction velocity is instantly updated (Figure 8.8).

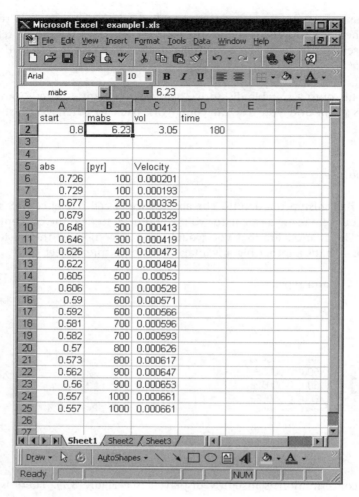

The spreadsheet shows the following data:

	A	B	C	D
1	start	mabs	vol	time
2	0.8	6.23	3.05	180
3				
4				
5	abs	[pyr]	Velocity	
6	0.726	100	0.000201	
7	0.729	100	0.000193	
8	0.677	200	0.000335	
9	0.679	200	0.000329	
10	0.648	300	0.000413	
11	0.646	300	0.000419	
12	0.626	400	0.000473	
13	0.622	400	0.000484	
14	0.605	500	0.00053	
15	0.606	500	0.000528	
16	0.59	600	0.000571	
17	0.592	600	0.000566	
18	0.581	700	0.000596	
19	0.582	700	0.000593	
20	0.57	800	0.000626	
21	0.573	800	0.000617	
22	0.562	900	0.000647	
23	0.56	900	0.000653	
24	0.557	1000	0.000661	
25	0.557	1000	0.000661	

Figure 8.8

Conclusions

You should be able to see now why a spreadsheet is so superior to using a calculator. You can see and edit all of the data and calculations so it is possible to correct mistakes, and the editing facilities make it easy to duplicate calculations.

Example 2   Graphing data

Excel can create graphs from your data. Unfortunately, from the vast range of graph styles most are business oriented. You can still produce a reasonable *x–y*

Example 2   Graphing data   169

graph though. Excel uses the word 'chart' for bar chart, pie chart etc. and for graphs also. A proper x–y type of graph is called a scatter chart. At the end of this example your graph will look like Figure 8.9.

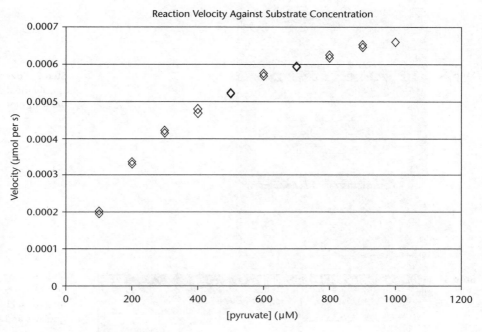

Figure 8.9   An Excel chart of the data.

You can plot a graph of the enzyme kinetics from the previous example. Detailed instructions are not provided here because Excel's Office Assistant will show you the steps. The following instructions will get you started.

▶ Open the file that you saved in Example 1. You need to select the concentration and reaction velocity data.

▶ Select cells B5 to C25 by depressing the mouse button on the centre of cell B5, dragging to cell C25 and releasing the button. Be careful to click in the middle of the cell because dragging the edge of the cell or the fill handle does different things.

▶ You are ready to create a graph on your spreadsheet. Choose Chart on the Insert menu.

▶ From now on follow the instructions that appear in the chart wizard. (Key dialogue boxes are shown in Figure 8.10.)

Step 1

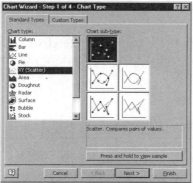

The first dialogue box allows you to select the style of the chart or graph

Step 2

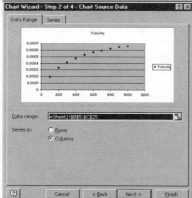

Step 2 allows you to specify the data you want to plot

Step 3

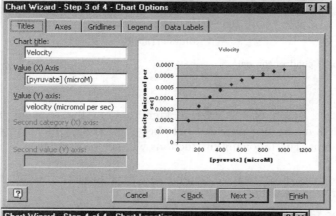

Step 3 allows you to add titles and select features such as gridlines and legends

Step 4

In step 4 you choose whether you want the chart placed on the sheet or inserted into the workbook

Figure 8.10

Example 3   Modelling the faecal occult blood test screening programme

In the case study on the faecal occult blood test you were looking at fictitious data because the real studies in the UK are yet to produce their results. Health policy makers don't have to sit idle while they wait for the results; they can start to think about the different policy decisions that they should make according to the different outcomes of the research. In other words, they can model the system mathematically and then study the model. What follows are instructions for implementing the model using Excel – you can look back at Chapter 7 for a discussion of the colorectal cancer screening programme.

The three inputs of the model are P(A), P(B|A) and P(B|NOT A). You can set these initially to the values given in the case study and alter them when the spreadsheet is complete to see what happens. Input these three figures and a number of text labels as shown in Table 8.1. Make sure you put your labels in the same cells as in the diagram otherwise the instructions that follow won't make sense.

Table 8.1

	A	B	C	D
1	Incidence			
2	P(A)	0.001		
3	P(NOT A)			
4				
5	With cancer			
6	P(B|A)	0.985		
7	P(NOT B|A)			
8				
9	Without cancer			
10	P(B|NOT A)	0.019		
11	P(NOT B|NOT A)			
12				
13	Population	A	NOT A	Total
14	B			
15	NOT B			
16	Total			
17				
18	Screening	P(A|B)	P(NOT A|B)	
19				
20		P(A|NOT B)	P(NOT A|NOT B)	
21				

You can now type the formulae shown in Table 8.2. You will see the same set of results on the sheet as was presented in the case study. The results are reproduced in Table 8.3.

Table 8.2

	A	B	C	D
1	Incidence			
2	P(A)	0.001		
3	P(NOT A)	=1-B2		
4				
5	With cancer			
6	P(B|A)	0.985		
7	P(NOT B|A)	=1-B6		
8				
9	Without cancer			
10	P(B|NOT A)	0.019		
11	P(NOT B|NOT A)	=1-B10		
12				
13	Population	A	NOT A	Total
14	B	=B6*B2	=B10*B3	=B14+C14
15	NOT B	=B7*B2	=B11*B3	=B15+C15
16	Total	=B14+B15	=C14+C15	
17				
18	Screening	P(A|B)	P(NOT A|B)	
19		=B14/D14	=C14/D14	
20		P(A|NOT B)	P(NOT A|NOT B)	
21		=B15/D15	=C15/D15	

Table 8.3

	A	B	C	D
1	Incidence			
2	P(A)	0.001		
3	P(NOT A)	0.999		
4				
5	With cancer			
6	P(B|A)	0.985		
7	P(NOT B|A)	0.015		
8				
9	Without cancer			
10	P(B|NOT A)	0.019		
11	P(NOT B|NOT A)	0.981		
12				
13	Population	A	NOT A	Total
14	B	0.000985	0.018981	0.019966
15	NOT B	0.000015	0.980019	0.980034
16	Total	0.001	0.999	
17				
18	Screening	P(A|B)	P(NOT A|B)	
19		0.049333868	0.950666132	
20		P(A|NOT B)	P(NOT A|NOT B)	
21		1.53056E–05	0.999984694	

Example 4 Modelling imprecision   173

You could argue that this has not saved you much time compared with doing the calculations using a pocket calculator, but you can do something with this spreadsheet that you couldn't do with a calculator. What if you wanted to know the implications of the disease being more common on the reliability of the screening programme. Just type a new figure, say 0.01 into cell B2. All of the calculations will recalculate instantly (Table 8.4).

Table 8.4

	A	B	C	D
1	Incidence			
2	P(A)	0.01		
3	P(NOT A)	0.99		
4				
5	With cancer			
6	P(B|A)	0.985		
7	P(NOT B|A)	0.015		
8				
9	Without cancer			
10	P(B|NOT A)	0.019		
11	P(NOT B|NOT A)	0.981		
12				
13	Population	A	NOT A	Total
14	B	0.00985	0.01881	0.02866
15	NOT B	0.00015	0.97119	0.97134
16	Total	0.01	0.99	
17				
18	Screening	P(A|B)	P(NOT A|B)	
19		0.343684578	0.656315422	
20		P(A|NOT B)	P(NOT A|NOT B)	
21		0.000154426	0.999845574	

The result is that many more of the positive results are true positives if the disease is more common, but at the same time more of the negatives are false negatives. You can appreciate that this type of mathematical modelling can be easy and a very powerful decision-making tool.

Example 4 Modelling imprecision

Mathematical modelling can be an educational tool as well as a decision-making aid. With complex mathematical systems you can test the properties of that system with the aid of models. As an example, instructions follow for modelling the string measuring study presented in Chapter 4. The discussion of the results is not duplicated here.

The complete model involves five sets of five replicate measurements. The total of 25 measurements will be calculated in cells N4 to N28, but before that stage we need the individual random elements of each measurement.

▶ In cell P4 enter the formula:

=IF(RAND( )<0.5,0.001,−0.001)

The function RAND( ) produces a random number between 0 and 1 with even distribution so the expression RAND( ) < 0.5 will be true 50% of the times it is recalculated. The IF(...) function makes sure that the entire formulae will produce either 0.001 or −0.001 as the result in the cell. If you press the F9 function key Excel will recalculate and the result of this calculation may change. This is one random event out of 140 for the first measurement.

▶ Duplicate the formula into cells Q4 to EY4. This is very easy. Select cell P4. You will see a small black box in the bottom right-hand corner of the cell which is called the fill handle. Just grab the fill handle with the mouse and drag it to the right. The spreadsheet will scroll and you just keep going until you reach EY4, at which point you can let go.

You now have 140 random contributions to the imprecision of one measurement which you need to total up.

▶ In cell O3 put the label:

R

▶ In cell O4 put the formula:

=SUM(P4:EY4)

▶ In cell N3 put the label:

X

▶ In cell N4 put the formula:

=42+O4

Now you have a single measurement and you need another 24. This is very easy.

▶ Start by selecting all the cells you have put formulae into in row 4, i.e. N4 to EY4.
▶ Drag the fill handle, which will be found in the bottom right of the selection, down to row 28.

All the calculations so far will be duplicated another 24 times and the results will be calculated instantly. You have the raw data and now you can go on to calculate some statistical values. Remember that we are putting measurements in groups of five, so the following statistics will apply to just the top five measurements.

Example 4   Modelling imprecision   175

▶ Enter the formulae shown in Table 8.5 to find the mean and standard deviation of your group of five replicate measurement.

There is a much quicker and easier way to calculate standard deviation but the method here breaks down the calculation into individual stages so you can more closely examine what's going on.

Table 8.5

	G	H	I	J	K	L	M
1							
2							
3	sd	var	SumSq	SqRes	Resid	Mean	SumX
4	=SQRT(H4)	=I4/4	=SUM(J4:J8)	=K4^2	=L4-N4	=M4/5	=SUM(N4:N8)
5				=K5^2	=L4-N5		
6				=K6^2	=L4-N6		
7				=K7^2	=L4-N7		
9				=K8^2	=L4-N8		

Now you need these calculations duplicated for the other four sets of five replicates each.

▶ Select the block of cells from G3 to M9.

▶ Select the Copy command on the Edit menu.

▶ Select cell G10.

▶ Select the Paste command on the Edit menu.

▶ Select cell G14, paste, select G19, paste, select G24, paste.

You now have five sets of sample statistics but it would be interesting to see how the sample standard deviations measure up to the population standard deviation. Because the mathematics behind this model are clearly understood it is possible to use a formula to determine the population standard deviation:

▶ In cell G1 type:

   =SQRT(0.5*0.5*140)*0.002

To gain insight into the meaning of standard error you can calculate it in two ways; for each group of replicates divide the sample standard deviation by the square root of five or, alternatively, calculate the standard deviation of the five sample means as if they were data points.

▶ Enter the formulae shown in Table 8.6.

▶ In cell A1 you can add the theoretical population standard error:

   =G1/SQRT(5)

Table 8.6

	A	B	C	D	E	F
1						
2						
3	sd	Sumsq	SqMeanRes	MeanRes	MeanMean	SE
4	=SQRT(B4/4)	=SUM(C4:C28)	=D4^2	=E4-L4	=SUM(L4:L28)/5	=G4/SQRT(5)
5						
6						
7						
8						
9			=D9^2	=E4-L9		=G9/SQRT(5)
10						
11						
12						
13						
14			=D14^2	=E4-L14		=G14/SQRT(5)
15						
16						
17						
18						
19			=D19^2	=E4-L19		=G19/SQRT(5)
20						
21						
22						
23						
24			=D24^2	=E4-L24		=G24/SQRT(5)

That's all the calculations completed and now you can get some insight on the random variation in the system by repeatedly recalculating the spreadsheet by pressing the F9 function key. Each time you recalculate, it's as if you have repeated the experiment. How much do the individual standard errors differ from the theoretical standard error and how much does the standard deviation of the five means differ from the theoretical standard error?

Example 5   The FOB test and horseradish peroxidase

For a long time, if you wanted to do a significance test with a computer, you would have to buy specialist statistical software, but these days ordinary spreadsheets like Excel have quite a wide range of statistical facilities. These facilities can be accessed in two ways: you can use statistical functions in your own formulae or you can select a command from the menu and get the spreadsheet to produce a whole set of statistical results in one go. You have to be very careful when you use

Example 5   The FOB test and horseradish peroxidase   **177**

statistical functions in Excel because sometimes the programmers have cut corners in the methods they use. For example, the function CHITEST which directly calculates a probability from frequencies, fails to apply the continuity correction that is needed when the degree of freedom is 1. The method described here side-steps the problem by calculating the χ^2 value using simple arithmetical formulae and then uses the CHIDIST function to find a probability. Table 8.7 shows exactly what you should enter in a spreadsheet to achieve the desired result.

▶ Enter the labels, data and formulae shown in Table 8.7 into a blank spreadsheet. The calculations are explained below.

Table 8.7

	A	B	C	D
1	OBSERVED	Control	Treatment	Total
2	Positive	15	21	=B2+C2
3	Negative	123	117	=B3+C3
4	Total	=B2+B3	=C2+C3	=D2+D3
5				
6	EXPECTED	Control	Treatment	
7	Positive	=B4*D2/D4	=C4*D2/D4	
8	Negative	=B4*D3/D4	=C4*D3/D4	
9				
10		=(ABS(B2-B7)-0.5)^2/B7	=(ABS(C2-C7)-0.5)^2/C7	
11		=(ABS(B3-B8)-0.5)^2/B8	=(ABS(C3-C8)-0.5)^2/C8	
12				
13	chi^2	=SUM(B10:C11)		
14	P	=CHIDIST(B13,1)		

Simple formulae with cell references are used to find subtotals and a total for the observed frequencies. These are quite self-explanatory. The expected frequencies each scale the column totals according to the proportion found from the appropriate row total and the grand total.

A 2×2 table calculates the value $(|O - E| - 0.5)^2/E$ for each observed frequency. You can see that the mathematical symbols don't naturally translate into the string of characters demanded by a spreadsheet program so it's worth pointing out the various elements. The value $|O - E|$ is found with ABS(B2–B7), the function ABS returning an absolute value. Spreadsheet formulae can't contain superscripted text and so ^2 is used to indicate a square. Finally the / symbol is always used for division because the formula must go on a single line of text.

The χ^2 value is found from the total of the values in the table, and that is what the SUM( ) function does. The cell reference in the SUM(B10:C11) function uses a colon, ':' to indicate a rectangular range of cells defined by its top left and bottom right corners. The CHIDIST function performs the complex calculation that finds p for the given χ^2 and degrees of freedom.

The strength of this spreadsheet is that it will perform a chi squared test on any 2×2 table of frequencies. You just replace the data and all the statistics will recalculate instantly.

Example 6   Simple assay validation

In this example instructions are given for creating a spreadsheet for estimating the precision and accuracy of an assay using replicate measurements of a reference standard solution. The spreadsheet could be used with any type of measurements in any units and the only assumptions made are that the data are normally distributed and that the value of the reference standard is known with perfect precision and accuracy.

Simply copy the labels and formulae of Table 8.8 into a blank spreadsheet.

Table 8.8

	A	B
1	Ref	100
2		
3	Count	=COUNT(B12:B111)
4	Mean	=AVERAGE(B12:B111)
5	S	=STDEV(B12:B111)
6	SE	=B5/SQRT(B3)
7	Deviation	=ABS(B4-B1)
8	T	=B7/B6
9	P	=TDIST(B8,B3-1,2)
10		
11	Data	in this column
12		100
13		110
14		120
15		130

A number of the formulae contain some of Excel's built-in statistical functions. COUNT, AVERAGE and STDEV all look at a range of cells and calculate the count, mean and sample standard deviation of the data, respectively. All these functions ignore cells in the given range that are empty so this spreadsheet will work with any number of data points up to one hundred. There isn't a built-in function for standard error, so for that value a formula including the standard deviation, the count and the square root function is used. The function TDIST is used to get a two-tailed probability from the t distribution. Excel has documentation on all these functions in the Help file. If all goes well you should see results that match Table 8.9.

Example 7 Hypothesis testing with a *t* test 179

Table 8.9

	A	B
1	Ref	100
2		
3	Count	4
4	Mean	115
5	S	12.90994449
6	SE	6.454972244
7	Deviation	15
8	T	2.323790008
9	P	0.102728079
10		
11	Data	in this column
12		100
13		110
14		120
15		130

Example 7 Hypothesis testing with a *t* test

Often, data are collected in order to test a hypothesis. In the assay validation situation the hypothesis was that the mean measurement deviated from the known value only because of imprecision. You can use a *t* test in an experimental situation as well as for this type of assay validation. Let's take a specific example. Imagine you have developed a new drug which is intended to reduce blood cholesterol and you want to find out if it works using volunteers. A simple experiment would be to take blood samples from the volunteers, give them a course of the drug and then take more blood samples. This type of experiment is called a paired test because for each individual there is a before and after measurement. We aren't really interested in the two separate cholesterol levels so much as the change in the level and so there is a single drop in cholesterol for each individual and those are the data we will analyse. The null hypothesis is that the mean change in blood cholesterol is non-zero owing to random variation between individuals only. This is a similar situation to Example 6 above; Table 8.10 highlights differences.

Despite the differences highlighted, in both situations a mean is tested against some reference point and the repeated measurements vary according to the normal distribution. That means that the same *t* test and the same spreadsheet can be used to analyse the data. You do need to calculate a drop in cholesterol for each individual though. The simplest solution would be to type before and after cholesterol measurements in columns C and D and put a formula in column B.

Table 8.10

Assay validation	Paired test
The reference concentration is mentioned in the null hypothesis	A drop in cholesterol of zero is mentioned in the null hypothesis
Tests the mean against the reference	Tests the mean drop in cholesterol against zero
Random variation occurs that is due to imprecision in the assay	Random variation occurs that is due partly to imprecision in the assay but due mainly to natural variation between individuals in the trial

Example 8   Block design assay validation

Often, scientists design assays for biomolecules and there is no way to obtain a standard reference solution for the substance involved. A block design attempts to detect inaccuracy in the assay protocol without the aid of a reference solution. The data from a block design consist of two or more blocks of replicate measurements, and it is the amount of variation between the block means that is important. An analysis of variance is the appropriate statistical method to apply to the data to find out if there is significant variation between block means. Excel has functions relating to AnoVa but there is a much simpler way to get Excel to do AnoVa without using formulae at all.

Before you do the statistics you have to make sure that the Analysis Tookpak Add-in is installed. Assuming that it is not installed at all, you will need to do the following;

▶ Run Microsoft Office Setup to add components to MS Office.

▶ Select Analysis Toolpak under the Excel group of software items.

When set-up is complete the appropriate files will be copied to your hard disk but you also need to tell Excel to use the add-in:

▶ Run Excel.

▶ Select Addins...on the Tools menu.

▶ In the dialogue box select Analysis Toolpak and then click on the OK button.

You now have an extra command, Data Analysis, on the Tools menu and you can do the statistics you need for this exercise.

▶ Start by typing your data into a blank spreadsheet (copy Table 8.11).

▶ Now select the Data Analysis command on the Tools menu. This will give you a dialogue box, shown in Figure 8.11, listing all the data analysis facilities that are available.

Example 8 Block design assay validation   181

Table 8.11

	A	B	C
1	2.15	2.22	2.08
2	2.21	2.19	2.14
3	2.19	2.15	2.09
4	2.18	2.21	2.12
5	2.21	2.19	2.12
6	2.22	2.17	2.12

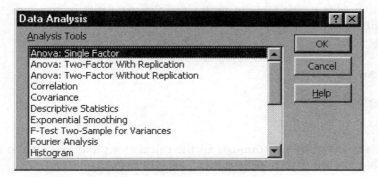

Figure 8.11

▶ Select AnoVa: Single Factor and click on the OK button. This will give you a dialogue box full of options. (See Figure 8.12.)

All the Data Analysis tools follow the same pattern. You have to tell the tool where the data are, where to put the analysis and select various options.

Figure 8.12

▶ Put the cursor in the input range control and then use the mouse to select on the spreadsheet the data you want to analyse. This should enter the appropriate cell reference in the control.

▶ The Grouped By control speaks for itself. Select Columns because the blocks are in columns.

▶ The Labels in First Row check-box tells the tool to use the labels in the top row of the data to label some of the statistical values. If you haven't used labels, leave this unchecked (i.e. unticked).

▶ Alpha refers to the critical probability. It makes little difference to the analysis but simply adds to the table of results the f statistic that corresponds to this probability.

▶ Select New Worksheet Ply and type

   my AnoVa

   into the box.

▶ Click on the OK button.

Excel will automatically run through all the calculations and present the results in a new ply (reproduced in Table 8.12).

Table 8.12

	A	B	C	D	E	F	G
1	SUMMARY						
2	*Groups*	*Count*	*Sum*	*Average*	*Variance*		
3	Column 1	6	13.16	2.193333	0.000667		
4	Column 2	6	13.13	2.188333	0.000657		
5	Column 3	6	12.67	2.111667	0.000497		
6							
7	ANOVA						
8	*Source of variation*	*SS*	*df*	*MS*	*F*	*P-value*	*F crit*
9	Between groups	0.025144	2	0.012572	20.72344	4.82E–05	3.682317
10	Within groups	0.0091	15	0.000607			
11							
12	Total	0.034244	17				

The table labelled Summary presents summary statistics separately for each of the blocks of data. This is followed by a table labelled AnoVa that contains the figures used for the significance test. For a full explanation of the statistical values produced by the AnoVa see page 93. The bottom line probability is 4.82×10^{-5} and this indicates a very highly significant variation between block means.

Example 9   Block design and hypothesis testing combined   183

Example 9   Block design and hypothesis testing combined

AnoVa can also be used for hypothesis testing in general. You can use it to compare two or more samples. Some drugs may result in cell damage and therefore elevated serum lactate dehydrogenase. So an experiment might look at serum lactate dehydrogenase in an experimental animal before and after dosing with a new drug. That experiment would produce two samples and a set of replicate measurements for each. AnoVa can be used to detect a significant difference between the before and after mean measurements. (Note that this is different from the paired test in Example 7 because we have replicate measurements of the before and after blood samples.)

In an experiment like the one just described it would be important to make consistently accurate measurements of lactate dehydrogenase and it's possible to combine hypothesis testing and block design for control of accuracy. So you might divide each of the blood samples into four portions and use four different centrifuges to spin down the cells in the blood sample. Replicate measurements are made on all the samples in the experiment using each of the blocks. In Table 8.13 both serum samples are assayed using each set of reagents. The results are set out in a table – there are eight sets of five replicates. (This experimental procedure goes a little beyond the normal amount of effort exerted in controlling accuracy – it would be more productive to spend the effort on using more experimental animals than assaying these two blood samples so many times.)

▶ Enter the data given in Table 8.13 into a blank spreadsheet.

Table 8.13

	A	B	C	D
Before	4.94	4.80	4.86	5.04
	5.19	4.96	4.82	4.85
	5.25	5.15	4.80	4.98
	5.15	5.08	5.20	4.93
	4.87	5.23	5.10	5.15
After	5.05	5.28	5.20	5.13
	5.30	4.91	5.21	5.32
	5.34	5.21	5.14	5.01
	5.28	5.16	5.16	5.23
	4.92	5.19	5.35	5.22

Two-factor AnoVa can simultaneously compare before and after samples and also compare blocks with each other. Once again it is only the probabilities produced that matter when it comes to drawing conclusions, but a lot of other figures are produced. As for one-factor AnoVa, a summary table is produced which gives basic statistics for each set of replicates. It also gives subtotals for blocks and samples.

▶ Select the Data Analysis command on the Tools menu and from the dialogue box select Two Factor AnoVa.

The options in the dialogue box are similar to the single-factor AnoVa and the results are presented in a similar layout. Table 8.14 shows the summary statistics and Table 8.15 shows the partitioning of variance, f and p values. The results are more complex than for a one-way analysis of variance. A discussion follows.

Table 8.14

	A	B	C	D	E	F
1		A	B	C	D	Total
2	Before					
3	Count	5	5	5	5	20
4	Sum	25.4	25.22	24.78	24.95	100.35
5	Average	5.08	5.044	4.956	4.99	20.07
6	Variance	0.0274	0.02843	0.03308	0.01285	0.10176
7						
8	After					
9	Count	5	5	5	5	20
10	Sum	25.89	25.75	26.06	25.91	103.61
11	Average	5.178	5.15	5.212	5.182	20.722
12	Variance	0.03362	0.01995	0.00677	0.01377	0.07411
13						
14	Total					
15	Count	10	10	10	10	
16	Sum	51.29	50.97	50.84	50.86	
17	Average	10.258	10.194	10.168	10.172	
18	Variance	0.06102	0.04838	0.03985	0.02662	

Table 8.15

	A	B	C	D	E	F	G
1	Source of variation	SS	df	MS	F	P-value	F crit
2	Sample	0.26569	1	0.26569	12.08575	0.001484	4.149086
3	Columns	0.01298	3	0.004327	0.196812	0.89779	2.901118
4	Interaction	0.04241	3	0.014137	0.643051	0.592996	2.901118
5	Within	0.70348	32	0.021984			
6	Total	1.02456	39				

Between samples

The mean for each sample is found. (There will be contributions from all blocks in each sample.) Variation between the sample means is calculated. This is owing to random variation plus any effect due to the drug that was used.

Example 10   Linear, non-weighted regression   185

Between blocks (columns)

The mean for each block is found. (All samples contribute to each mean.) Variation between the block means is calculated. This is owing to random variation plus any effect due to inaccuracy.

Interaction

This is owing to random variation plus interaction between sample and block. This tells you if the two samples had significantly different inaccuracy. It's difficult to see how this could happen with the block design used but it's still worth checking.

Within replicates

A mean is calculated for each individual set of replicates. This is exactly the same as for one term AnoVa. It is due entirely to imprecision.

Total

The total sums of squares is the sum of the four contributing sums of squares. (Or it could be calculated by taking all of the assay results together as if they were one big set of replicates.)

Tests

There is more than one test to perform. In fact there are three questions to answer for your data:

▶ Is there a significant difference between samples? The p value is 0.001484 which is well below 0.05 so the different sample means didn't occur by chance – there is a highly significant difference. The drug caused raised serum lactate dehydrogenase.

▶ Is there a significant difference between blocks? The p value is 0.89779 which is well above 0.05 so the different block means could easily have occurred because of imprecision alone. No inaccuracy was found.

▶ Is there a significant interaction between block and sample? The p value is 0.592996 which is well above 0.05 so the observed interaction could easily have occurred by chance alone. There is no interaction between sample and block. Accuracy was equal in both serum samples.

Example 10   Linear, non-weighted regression

This example shows you how to get Excel to apply a non-weighted linear regression to data. The data are taken from the lactate dehydrogenase kinetics study

Table 8.16

	A	B	C	D
1	[S]	V0	1/[S]	1/V0
2	(micM)	micromol/s	1/micM	s/micromol
3	100	0.0101		
4	100	0.01		
5	100	0.0097		
6	100	0.0104		
7	200	0.0165		
8	200	0.0167		
9	200	0.0164		
10	200	0.0166		
11	300	0.0209		
12	300	0.0207		
13	300	0.021		
14	300	0.0207		
15	400	0.0238		
16	400	0.0236		
17	400	0.0242		
18	400	0.0242		
19	500	0.0266		
20	500	0.0265		
21	500	0.0266		
22	500	0.0262		
23	600	0.0283		
24	600	0.0285		
25	600	0.0283		
26	600	0.0285		
27	700	0.0298		
28	700	0.0298		
29	700	0.0297		
30	700	0.0294		
31	800	0.0312		
32	800	0.0314		
33	800	0.0313		
34	800	0.0309		
35	900	0.0322		
36	900	0.0324		
37	900	0.0326		
38	900	0.0326		
39	1000	0.0334		
40	1000	0.0331		
41	1000	0.033		
42	1000	0.0332		

Example 10   Linear, non-weighted regression   187

described in Chapter 6 and you should remember that a double reciprocal plot of substrate concentration against initial reaction velocity gives a straight line although it is not heteroschedastic, and therefore non-weighted linear regression is not strictly valid. The instructions below also show you how to make a residual plot from the regression, which will reveal the obvious heteroschedastic nature of the transformed data.

▶ Copy the data shown in Table 8.16 into a blank spreadsheet.

▶ Add the formula

=1/A3

to cell C3 and the formula

=1/B3

in cell D3.

▶ Select C3 and D3 and then use the fill handle to duplicate the formulae down to row 42. This gives you the transformed data.

▶ Select Data Analysis from the Tools menu.

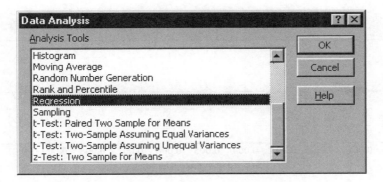

Figure 8.13

▶ In the dialogue box, shown in Figure 8.13, select Regression and click on OK. This will give a dialogue box full of regression options.

▶ Fill in the dialogue box as in Figure 8.14. (Note that the box asks for Y values before it asks for X values, which is a bit odd.)

The new ply (layer) in the spreadsheet created by this tool was given in the earlier chapter but Figure 8.15 shows the residual plot again.

You can see right away that the data are heteroschedastic but that it isn't quite as certain that they are linear. It's only possible to detect serious problems by eye.

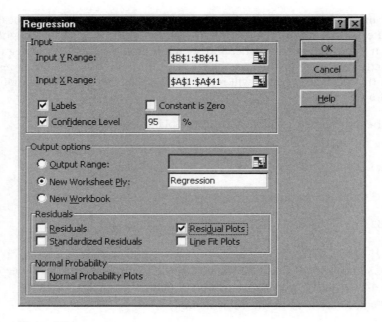

Figure 8.14

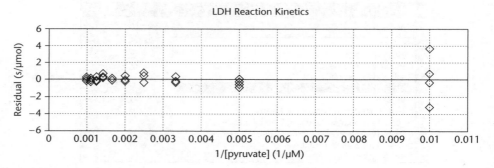

Figure 8.15   Double reciprocal residual plot created using Excel.

Example 11   Regression – Testing for non-linearity

In this example you can take the spreadsheet you created in Example 10 and add a test for non-linearity.

▶ Copy the transformed data, in blocks, to sheet 2 of your spreadsheet and use the reciprocal concentrations as labels. The idea is to get replicate measurements in blocks suitable for AnoVa as shown in Table 8.17.

▶ Select Data Analysis from the Tools menu and select Single Factor AnoVa.

▶ Fill in the dialogue box exactly as in Figure 8.16. (It is particularly important that you select the Labels in First Row button.)

Example 11   Regression – Testing for non-linearity   189

Table 8.17

	A	B	C	D	E
1	0.01	0.005	0.003333	0.0025	Etc.
2	99.0099	60.60606	47.84689	42.01681	
3	100	59.88024	48.30918	42.37288	
4	103.0928	60.97561	47.61905	41.32231	
5	96.15385	60.24096	48.30918	41.32231	

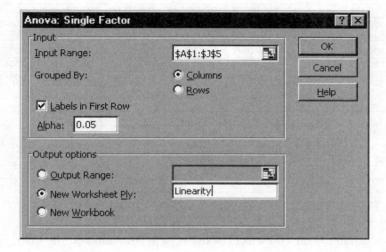

Figure 8.16

Table 8.18 is the output of the analysis tool with two columns, labelled Predicted and Meansq, added to the summary statistics. (Instructions follow.)

▶ The formula in cell F5 is

=A5*Regression!$B$18+Regression!$B$17

Note that this formula uses a type of cell reference that refers to one of the other sheets, Regression in the spreadsheet.

▶ The formula in cell G5 is

=(D5-F5)^2/B5

This is the mean squared deviation of the block mean from the value predicted by the regression.

▶ Select cells F5 and G5 and use the fill handle to duplicate the two formulae.

At this point you need to break down the 'between groups' partition of variance and this requires some extra cells in the AnoVa table.

Table 8.18

	A	B	C	D	E	F	G
1							
2							
3	SUMMARY						
4	*Groups*	*Count*	*Sum*	*Average*	*Variance*	Predicted	Meansq
5	0.01	4	398.25653	99.564133	8.1928613	99.37701	0.008753
6	0.005	4	241.70287	60.425718	0.2221951	60.88879	0.053609
7	0.003333333	4	192.0843	48.021074	0.1193247	48.05938	0.000367
8	0.0025	4	167.03432	41.758579	0.2749011	41.64468	0.003243
9	0.002	4	151.09176	37.772939	0.0738166	37.79586	0.000131
10	0.001666667	4	140.84682	35.211704	0.0204963	35.22998	8.35E–05
11	0.001428571	4	134.79773	33.699433	0.0467054	33.3972	0.022836
12	0.00125	4	128.20976	32.052439	0.0496628	32.02262	0.000222
13	0.001111111	4	123.26979	30.817448	0.0332385	30.95351	0.004628
14	0.001	4	120.57511	30.143778	0.0239881	30.09821	0.000519
15							
16							
17	ANOVA						
18	*Source of variation*	*SS*	*df*	*MS*	*F*	*P-value*	*F crit*
19	Between groups	16400.152	9	1822.2391	2011.9255	3.44E–39	2.210697
20	Within groups	27.171569	30	0.905719			
21							
22	Total	16427.324	39				

▶ Select cells A20 to G21 and select the Insert Cells command on the Insert menu. When asked, select the option to move the other cells down.

▶ Enter the formulae shown in Table 8.19 in the cells of the new rows.

Table 8.19

	A	B	C	D	E	F	G
1	ANOVA						
2	*Source of variation*	*SS*	*df*	*MS*	*F*	*P-value*	*F crit*
3	Between groups	16400.1522	9	1822.23914	2011.92553	3.4401452E–39	2.2106974
4	**Regression**	=B19-B21	1	=B20/C20			
5	**Non-linearity**	=SUM(I5:I14)	9	=B21/C21	=D21/D22	=FDIST(E21,C21,C22)	
6	Within groups	27.171569	30	0.90571898			
7							
8	Total	16427.3238	39				

This should give the results that are shown in Table 8.20.

To the limit of the number of significant figures used by Excel the probability of the null hypothesis is 1. That means that there is no evidence of non-linearity at all.

Example 12   Regression – Testing for heteroschedasticity   191

Table 8.20

	A	B	C	D	E	F	G
1	ANOVA						
2	*Source of variation*	*SS*	*df*	*MS*	*F*	*P-value*	*F crit*
3	Between groups	16400.15	9	1822.239	2011.926	3.44E–39	2.210697
4	Regression	16400.06	1	16400.06			
5	Non-linearity	0.094392	9	0.010488	0.01158	1	
6	Within groups	27.17157	30	0.905719			
7							
8	Total	16427.32	39				

Example 12   Regression – Testing for heteroschedasticity

In this example we add a few calculations to the spreadsheet produced in the previous two examples to test for heteroschedasticity.

▶ Starting from cell B26 enter the labels and formulae given in Table 8.21.

Table 8.21

	B	C	D	E
26	s^2max	s^2min	Fmax	p
27	=MAX(E5:E14)	=MIN(E5:E14)	=B27/C27	=FDIST(D27,2,2)*2

This should give the results shown in Table 8.22.

Table 8.22

	B	C	D	E
26	s^2max	s^2min	Fmax	p
27	8.192861	0.0204963	399.7233	0.004991

The probability is way below the conventional critical probability of 0.05 (i.e. greater than 95% significance) and so we have to reject non-weighted, non-linear regression as an invalid statistical method in this case.

Example 13   Non-linear, non-weighted regression

Excel '97 is not capable of performing the iterative line-fitting methods and another software product is needed. It's possible to obtain third-party add-ins to

Excel that perform more advanced statistical functions but this example uses a product called Origin version 5.0, produced by MicroCal. The idea is to show how the user interface works for a typical statistical package, and instructions may be different in their detail for other software products. Origin uses a concept similar to a spreadsheet for data entry, and when you start with a new spreadsheet there are two columns of cells to enter data into.

▶ Type the substrate concentration and reaction velocity data from Table 8.16 into columns A and B of the worksheet as shown in Figure 8.17.

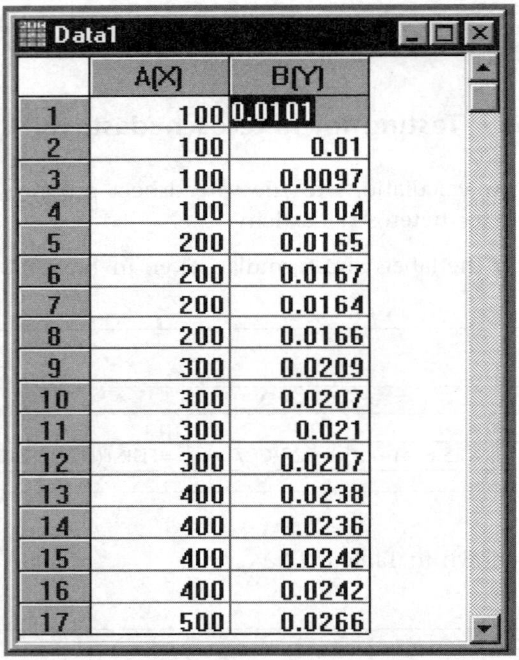

Data1	A(X)	B(Y)
1	100	0.0101
2	100	0.01
3	100	0.0097
4	100	0.0104
5	200	0.0165
6	200	0.0167
7	200	0.0164
8	200	0.0166
9	300	0.0209
10	300	0.0207
11	300	0.021
12	300	0.0207
13	400	0.0238
14	400	0.0236
15	400	0.0242
16	400	0.0242
17	500	0.0266

Figure 8.17

▶ Choose the command Non-Linear Curve Fit...on the Analysis menu. This will bring up a complex dialogue box (Figure 8.18). Actually there are several different dialogue boxes you can switch between using buttons.

The first stage is to make sure that the software knows the equation for the curve that defines the relationship between substrate concentration and reaction velocity:

▶ From the Functions list box select Hyperbl. When it is selected you should see a picture of the equation which is familiar from Michaelis–Menton reaction kinetics.

▶ Now press the Select Dataset button to obtain the second dialogue box (Figure 8.19).

Example 13   Non-linear, non-weighted regression   193

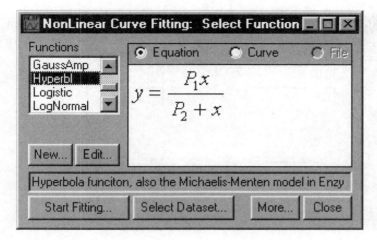

Figure 8.18

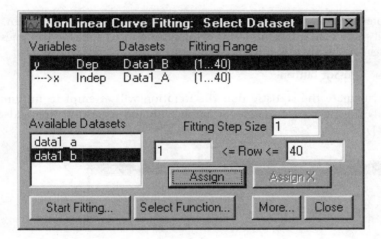

Figure 8.19

▶ In this dialogue box select first select *y* in the Variables list box then data1_b in the Available Datasets list box and connect the two by pressing the Assign button.

▶ Do the same thing to connect the *x* variable with data1_a. (Use the Assign button; the Assign X button is not appropriate for this example.)

▶ Now press the Start Fitting…button to get the final dialogue box (Figure 8.20).

In this dialogue box P1 refers to $V_{\max}$ and P2 refers to K_m. You can set these to any starting values you like.

▶ Enter first-guess values for P1 and P2. Say P1 = 0.02 and P2 = 200.

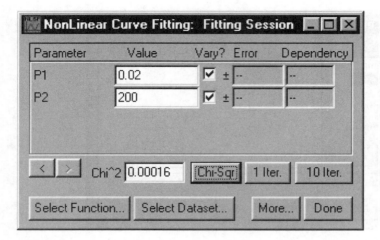

Figure 8.20   In this dialogue box P1 refers to V_{max} and P2 refers to K_m.

(Reasonable first guesses would be V_{max} = highest velocity in experiment, K_m = middle substrate concentration, but it is interesting to set them to bad guesses so you can see the iterative algorithm doing more work to improve the fit.)

▶ Press the Chi-Sqr button.

This will calculate the statistic that the iteration will attempt to minimize. It also signals the software to redraw a graph with the regression line added (see Figure 8.21).

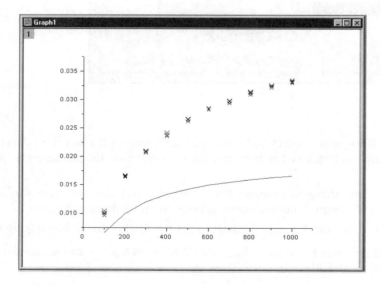

Figure 8.21

Example 13 Non-linear, non-weighted regression 195

Repeatedly press the 1 Iter button and watch the coefficients change as each iteration is performed. You should see the χ^2 statistic consistently decrease as the regression line gets closer and closer to the data, but eventually pressing the 1 Iter button will no longer cause any improvement in the statistical values. The graph will be updated also (see Figure 8.22). At this point you can make a note of the final parameters found. They're shown in Figure 8.23.

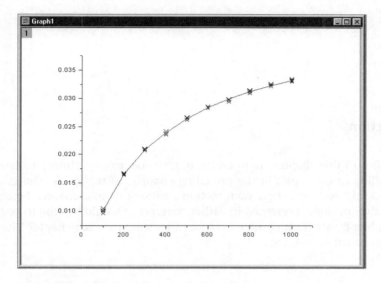

Figure 8.22

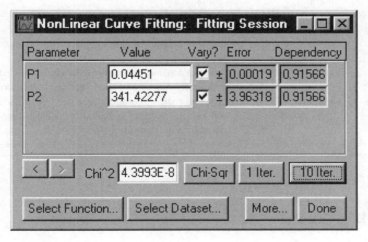

Figure 8.23

Chapter 9

Exercises

Introduction

The exercises in this chapter are intended to give you an opportunity to test your understanding of the topics in the preceding chapters of this book. The exercises are divided into sections where each section corresponds to a chapter. Since each chapter builds on ideas presented in earlier chapters, it would be wise to test your understanding by attempting some exercises after studying each chapter. The exercises are classified as follows:

Symbol	Meaning	Explanation
✍	Quick calculation	Involves simple arithmetic or algebra
⧖	Time-consuming calculations	Involves a series of calculations and/or algebra that will take more time
↝	Interpretation	Involves looking at experimental situations or the results of analysis and then reaching conclusions
🖱	Computer	Involves making computations using a spreadsheet
🗁	Open ended	An exercise with no specific right answer that tests your communication or decision-making skills

Probability

Exercise 1

Table 9.1 contains probability figures given using three different conventions. Fill in the blanks using the information given.

Table 9.1		
Mathematical	**Percentage**	**Integer ratio**
		1 in 3
	25%	
0.01		
	0.01%	

Exercise 2 ✍

In a study run in Ruritania, in the period 1980–82, records were kept on the number of babies born with hypothyroidism. The results obtained are shown in Table 9.2.

Table 9.2	
Total number of babies born during the period of study	**Number of babies born with hypothyroidism**
210,125	35

According to the data, what is the probability that a baby born in Ruritania will have hypothyroidism? (Round to 3 significant figures.)

Exercise 3 ✍

In another study run in the republic of Santa Cruz in 1990–92, records were again kept on numbers of babies born with hypothyroidism but this time results were recorded separately for boys and girls. The results are shown in Table 9.3.

Table 9.3		
	Total number of babies born during the period of study	**Number of babies born with hypothyroidism**
Male	153,205	17
Female	152,800	33

According to the data, what is the probability that a girl in the republic of Santa Cruz will be born with hypothyroidism? What is the probability that a boy in the republic of Santa Cruz will be born with hypothyroidism? (Round to 3 significant figures.)

Exercise 4 📁

The probabilities found in Exercises 2 and 3 will be applied to the projected birth rate in order to predict the number of babies that will need treatment for hypothyroidism over the next year. Was the study in Exercise 3 better than the study in Exercise 2 for this purpose? Did the study in Exercise 2 produce a valid probability? Were the probabilities in Exercise 3 equally valid?

Do the answers to these questions change if the aim of the study is to give information to pregnant women about the risk that their unborn child will be born with hypothyroidism? If you think that one or more of the probabilities found is invalid for either purpose, what should health authorities do when they plan for the next year?

To what extent do you think health authorities in Ruritania should depend on data from Santa Cruz, and vice versa?

Exercise 5 ✍

In an experiment to test a blood-pressure drug, the blood pressure of twenty rats was measured before and after dosing. A statistical test was used to determine the probability that the mean drop in blood pressure was due only to ordinary fluctuations in blood pressure, $P(H_0)$ and not a result of the drug.

$$P(H_0) = 2.5\%$$

Calculate the probability that the mean drop in blood pressure is significant, $P(\text{NOT } H_0)$.

Exercise 6 ✍

A screening programme for colorectal cancer is being planned using a type of faecal occult blood test. A study has determined the probabilities of people falling into one of four categories. The results are presented in Table 9.4.

Table 9.4	
Group	**Probability**
True positive result	0.001%
False positive result	0.999%
True negative result	98.901%
False negative result	0.099%

(a) What is the probability that a person will get a positive result with this faecal occult blood test?

(b) What is the probability that a person will get a negative result with this faecal occult blood test?

Exercise 7 ⏳

Imagine you are a health worker at a university student health clinic. You want to advise women who are taking medication based on steroids (e.g. the contraceptive pill) and are regular smokers that they are at high risk of certain health problems and should give up smoking. For simplicity, you would like your receptionist to hand out pamphlets to all women entering the clinic without asking them any questions. This would certainly reach the group in question but may be inefficient. To decide if this is a reasonable strategy you need to know what the probability is that a woman entering the clinic is taking steroid-based drugs *and* is a smoker.

In a survey of medical records (performed with the permission of patients), 205 female patients were taking steroid-based drugs out of a sample of 675.

(a) Based on these data, what is the probability that a woman entering the clinic is taking steroid-based drugs?

(b) Conversely, what is the probability that a woman entering the clinic is *not* taking steroid-based drugs?

In a separate survey carried out by reception staff, women entering the clinic were asked if they regularly smoke. Of 702 women asked, 278 said they were regular smokers.

(c) Based on these data, what is the probability that a woman entering the clinic is a regular smoker?

(d) Conversely, what is the probability that a woman entering the clinic is *not* a regular smoker?

These probabilities can be combined to determine the following. What is the probability that a woman patient is:

(e) Taking steroid drugs and is a smoker?

(f) Taking steroid drugs and is not a smoker?

(g) Not taking steroid drugs and is a smoker?

(h) Not taking steroid drugs and is not a smoker?

(i) In what ways is this experimental design imperfect? How would you do a better study?

Exercise 8 ⏳

This is quite a long and involved exercise but the mathematics has many similarities with the case study on the faecal occult blood test presented in Chapter 7 so if you get stuck revise that chapter.

There is a particular 'over the counter' drug used to clear nasal congestion. It is reported that there is a health risk connected with people who are on a certain medication for high blood pressure and who use this nasal decongestant regularly because the decongestant interacts with the drug and makes it less effective. There is need for a study which can be used by policymakers to decide how to tackle the issue. You must be interested in people who take and who do not take the blood-pressure drug and in people who use or do not use the nasal decongestant. The most important statistic to determine is the probability that a person taken from the population at random is taking the blood-pressure drug and using the nasal decongestant.

A letter was sent to a random sample of 1032 people from the general population asking them whether they were using the drug and whether they were using the nasal spray (results in Table 9.5). Unfortunately, no one in the sample was taking the blood-pressure drug so the data only tell us about people who don't take the drug.

Table 9.5

Observed	Not taking drug
Use nasal spray	100
Don't use spray	932
Total	1032

In a second stage to the study, medical records at a number of clinics were used to identify patients taking the drug. Only those using the drug were asked if they use the nasal spray in question. Of the 98 456 medical records checked, 354 patients were found to be taking the blood-pressure drug.

The researchers managed to contact 256 of the 354 users of the blood-pressure drug by telephone and discovered that 55 use the nasal spray and 201 do not use the nasal spray.

(a) From the medical records, what is the probability that a person taken at random from the population is taking the blood-pressure drug? **P(A)**

(b) From the medical records, what is the probability that a person is *not* taking the blood-pressure drug? **P(NOT A)**

(c) From the telephone interviews, what is the probability that a blood-pressure drug user also uses the nasal spray? **P(B|A)**

(d) From the telephone interviews, what is the probability that a blood-pressure drug user does *not* use the nasal spray? **P(NOT B|A)**

(e) From the letters sent to homes, what is the probability that a non-user of the blood-pressure drug uses the nasal spray? **P(B|NOT A)**

(f) From the letters sent to homes, what is the probability that a non-user of the blood-pressure drug does *not* use the nasal spray either? **P(NOT B|NOT A)**

As a general point, are users of the blood-pressure drug more likely to be using the nasal spray?

Perhaps the drug is causing nasal congestion as a side effect. Calculate the following:

(g) What is the probability that a person taken at random from the population is taking the blood-pressure drug and using the nasal spray? P(A AND B)

(h) What is the probability that a person taken at random from the population is taking the blood-pressure drug and *not* using the nasal spray? P(A AND NOT B)

(i) What is the probability that a person taken at random from the population is *not* taking the blood-pressure drug and using the nasal spray? P(NOT A AND B)

(j) What is the probability that a person taken at random from the population is *not* taking the blood-pressure drug and *not* using the nasal spray? P(NOT A AND NOT B)

Exercise 9

A certain operation carries some risk of death. Records were kept in a number of hospitals that specialize in this surgical procedure over a period of several years. The records show that the procedure was carried out 7456 times and on 23 occasions the patient died during or soon after the operation. So the probability that a patient will die as a result of this operation is 0.31%. If you were a surgeon in this hospital and the father of a patient asked you, 'How safe is this operation?', how would you reply? Choose your words carefully and make no assumptions about the educational background of the parent.

Exercise 10

Imagine that the patient referred to in the previous exercise died during the operation. The father of the patient says to you, 'I thought this operation was safe but my son died. I'm going to sue the hospital over this.' How will you reply to this? Do you think your response under Exercise 9 could have helped to avoid this situation occurring?

Significance

Exercise 11

In Chapter 2 an experiment was presented that was intended to show if horseradish sauce causes false positives for the faecal occult blood test. Imagine that the experiment did indicate a significant increase in positive results when horseradish sauce was included in the diet. Design an experiment that would distinguish the hypothesis that false positives are caused by peroxidase in the horseradish from the hypothesis that false positives are caused by some other ingredient in the horseradish sauce.

Exercise 12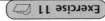

Table 9.6 summarizes results from Exercise 8.

Table 9.6

Observed	Using drug	Not taking drug	Total
Using spray	55	100	155
Not using spray	201	932	1133
Total	256	1032	1288

What are the expected frequencies? Fill in Table 9.7. (If necessary review the meaning of the phrase 'expected frequencies' in Chapter 2. N.B. The expected frequencies in this exercise need not be whole numbers.)

Table 9.7

Expected	Using drug	Not taking drug	Total
Using spray			155
Not using spray			1133
Total	256	1032	1288

Exercise 13 ⧗

Take the frequencies from the previous exercise and calculate χ^2. Use a pocket calculator to implement the method described at the end of Chapter 2.

Exercise 14 🖑

Repeat Exercise 13 using a computer. This time go further than calculating χ^2 and calculate the probability of the null hypothesis too.

Measurements and assays

Exercise 15 🖑

In Chapter 3 the lactate dehydrogenase assay was described. Design a spreadsheet so that you can type light absorbance readings into part of the spreadsheet and see the activity of the enzyme calculated in another.

Exercise 16 ᘗ

Here are four values with units:

0.05 (absorbance change – no units)

3.0 ml

5.0 ml µmol⁻¹ cm⁻¹

1 cm

The calculations in Table 9.8 are attempts at a calculation using these four figures. For each calculation work out the units for the result. Don't bother calculating the actual result, just decide on the units the number will have.

Table 9.8

Calculation	Units
$\dfrac{0.05 \times 3.0}{5.0 \times 1.0}$	
$\dfrac{5.0 \times 3.0}{0.05 \times 1.0}$	
$\dfrac{0.05 \times 1.0}{5.0 \times 3.0}$	
$\dfrac{0.05 \times 3.0 \times 1.0}{5.0}$	

Exercise 17 ✎

In Table 9.9 are the results for the construction of a calibration curve from an immunoassay for α-foetoprotein. The details of the assay are not important but the basic principle is as follows. The radioactivity in a mixture of reagents in a tube is measured. A further reagent is added that causes precipitation of protein and the radioactivity bound to the precipitated protein is measured. The percentage in the table is the amount of bound radioactivity in the precipitate as a percentage of the radioactivity in the tube. The concentration of the standard solution that was used in the experiment is shown in the first column.

Table 9.9

ng ml⁻¹	Bound (%)
0.25	11.8
0.5	10
1	8
2.5	5
5	3.5
10	2

(a) Plot the data with concentration on the x-axis and radioactivity bound (%) on the y-axis and then fit a calibration curve to the points.

(b) A blood serum sample was assayed for α-foetoprotein and the result was 8.5 bound (%). What is the concentration of α-foetoprotein in the blood serum sample?

Exercise 18 ✍

A laboratory made five replicate measurements of a standard solution of insulin. Their mean was very significantly higher than the correct value. The method was altered and accuracy was very much improved, without improving the precision, and five more measurements made. Which of the following statistical values will be lower in the second set of data compared with the first?

(a) The mean.

(b) The lower confidence limit of the mean.

(c) The upper confidence limit of the mean.

(d) The standard error of the mean.

(e) The standard deviation.

Exercise 19 ✍

Convert the following numbers to scientific notation:

(a) 3127

(b) 0.00791

Evaluate the following and give the result in scientific notation:

(c) $6.023 \times 10^{23} \times 1.0 \times 10^{-27}$

(d) $(2.0 \times 10^{-5} \times 4.2 \times 10^{2}) + 4.5 \times 10^{-4}$

(e) $3.054 \times 10^{-6} + 5.78 \times 10^{-7}$

(f) $2.23 \times 10^{16} - 8.8 \times 10^{14}$

(g) $\dfrac{1.2 \times 10^{-1}}{4.0 \times 10^{-2}}$

(h) $\dfrac{3.0 \times 10^{-5} + 5.5 \times 10^{3}}{5.0 \times 10^{17}}$

Exercise 20 ✍

Simplify each of the following:

(a) $5^4 \times 5^2$

(b) $(3^3)^2$

(c) $\dfrac{x^3 x^2}{x^4}$

(d) $\dfrac{3x^2}{x^5x^3}$

(e) $\dfrac{y^2x^5}{x^2}$

(f) $\left(\sqrt[3]{x}\right)^6$

(g) $\dfrac{x^2\sqrt[2]{x}}{x^{3.5}}$

Exercise 21 ✍

Find x in each of the following expressions without using a pocket calculator:

(a) $\log_3 27 = x$

(b) $\log_{16} x = \frac{1}{2}$

(c) $\log_x 8 = \frac{1}{2}$

Rearrange the following for x:

(d) $y = \log_b x$

(e) $y = \log_b(x^3)$

(f) $y = \log_x 3$

(g) $y = \log_b\left(\sqrt{x^3}\right)$

With an unknown base, b, here are two logarithms:

$\log_b 2 = 0.69$ and $\log_b 3 = 1.1$

Without using a pocket calculator, use the logs given to evaluate the following:

(h) $\log_b 6$

(i) $\log_b\left[3 \times \sqrt[2]{3}\right]$

(j) $\log_b\left[12^{1/4}\right]$

Another unknown base, c, give the following two logarithms:

$\log_c 3 = 0.95$ and $\log_c 4 = 1.20$

Once again evaluate the following without the aid of a calculator:

(k) $\log_c 9$

(l) $\log_c 16$

(m) $\log_c 12$

(n) $\log_c(3^3 \times 4^6)$

Precision

Exercise 22 ✍

The α-foetoprotein assay produces a percentage of bound radioactivity (See Exercise 17 above). What are the 95% upper and lower confidence limits of a single measurement of 10 bound (%), assuming that the standard deviation is 0.2 bound (%)?

Exercise 23 ⌛

The following five replicate measurements for α-foetoprotein were made on a serum sample:

Bound (%)

4.8

4.7

5.2

5.1

5.1

Use a pocket calculator to calculate the mean and sample standard deviation of these data.

Exercise 24 ✍

(a) Find the mean and sample standard deviation of the following eleven replicate measurements of α-foetoprotein in a serum sample:

% bound radioactivity
3.56   3.42   3.40   3.36   3.51   3.43   3.56   3.57   3.46   3.58   3.44

(b) Now find the standard error of the data.
If you were to use a computer or statistical table you would find that $t_{0.05} = 2.228$ when $(N - 1) = 10$.

(c) Find the upper 95% confidence limit of the mean.

(d) Find the lower 95% confidence limit of the mean.

(e) Now state your measurement in the following form:

Estimated concentration = _____ bound (%). The true value lies between _____ bound (%) and _____ bound (%) with 95% confidence.

Exercise 25 ⌛

This exercise involves using the α-foetoprotein calibration curve you produced in Exercise 17 above.

Imagine you have a serum sample which produced a value for bound radioactivity of 6%, and upper and lower confidence limits are 6.5% and 5.5%, respectively. Estimate the α-foetoprotein concentration using your calibration curve. Find the 95% confidence limits in ng ml^{-1} by constructing lines on the calibration curve (i.e. exactly as if you were treating 6.5% and 5.5% as measurements you were looking up).

A second serum sample produced a value for bound radioactivity of 3%, and upper and lower confidence limits are 3.5% and 2.5%, respectively. This is equal precision to the previous measurement. Repeat the same procedure to convert these figures into α-foetoprotein concentration in ng ml^{-1}.

For this assay, determination of percentage-bound radioactivity has constant precision for high and low values. When the calibration curve has been used to determine concentrations in mass per unit volume, is precision still constant for high and low concentrations? Do you think variation in the estimation of concentration in mass per unit volume is likely to be normally distributed?

Exercise 26 🗁

A pregnant woman carrying a normal foetus will have an elevated concentration of α-foetoprotein in her blood serum compared with before the pregnancy. If a pregnant woman is carrying a foetus with spina bifida this concentration rises even more. If the laboratory in a hospital was producing assay results for α-foetoprotein that were much more imprecise than could be expected, what could the implications be for prenatal screening for spina bifida?

Accuracy

Exercise 27 〜

The α-foetoprotein assay can become inaccurate because the radiolabelled α-foetoprotein used in the assay may adhere to the sides of the container. (This is avoided by using glass tubes and raising the total protein concentration with bovine serum albumin.) Five different laboratories were given the same standard solution of α-foetoprotein to assay using five replicate measurements. The summary statistics are in Table 9.10. Reference standard = 5.20 bound (%).

Table 9.10					
Lab	Mean	SE	dev.	t	p
1	5.18	0.48	0.02	0.042	0.969
2	4.33	0.45	0.87	1.933	0.125
3	5.19	0.22	0.01	0.045	0.966
4	4.51	0.26	0.69	2.654	0.057
5	5.11	0.21	0.09	0.429	0.690

(a) Rank the five laboratories according to amount of imprecision in their assay.

(b) Rank the five laboratories according to the amount of inaccuracy in their assay.

(c) Rank the five laboratories according to the significance of inaccuracy in their assay.

Exercise 28 ⏳

A standard solution of lactate dehydrogenase was repeatedly assayed. Find out whether the mean deviates significantly from the known value. Reference standard = 4 μmol s^{-1} l^{-1}.

The replicate measurements given in Table 9.11 are in the same units.

Table 9.11		
3.90	4.10	3.92
4.15	3.85	3.99
3.70	4.10	4.12
4.25	4.18	

(a) Calculate a mean.

(b) Calculate the standard error of the mean.

(c) Calculate the deviation of the mean from the reference standard.

(d) Calculate the t statistic – divide the deviation by the standard error.

The next step would normally be to calculate the probability of the null hypothesis. The null hypothesis is that the mean measurement differs from the reference standard only because of the imprecision of the assay. In other words it is the probability that $\bar{X} - x = \bar{R}$. If you don't have a computer you can't do that but you can at least find out if the probability is greater or less than 0.05.

(e) Given $t_{0.05} = 2.228$ when $(N - 1) = 10$, calculate 95% confidence limits on the reference standard. Does your mean measurement lies outside or inside the 95% confidence limits? What does this mean?

Exercise 29 🖱

(a) From the previous exercise, calculate the probability of the null hypothesis. You won't be able to do this with your pocket calculator so use a computer. Calculate $1 - p$ also.

(b) Complete this statement:
The measured concentration differs from the reference standard by _____ μmol s^{-1} l^{-1}. The probability that this is because of random variation in the individual measurements alone is _____. The probability that it is owing partly to inaccuracy is _____.

Exercise 30 🐛

A certain laboratory is making measurements of α-foetoprotein in blood samples and the experiment requires that estimates are very accurate and precise. Consequently, a large number of replicates are made on each blood sample and these are performed in three batches using separate reagents and equipment.

The results for the assay of a single blood serum sample were analysed with AnoVa. Examine the printout from the spreadsheet in Table 9.12. Decide if there is significant variation between groups of measurements (using $p = 0.05$ as your critical probability). If it's appropriate, calculate the overall mean.

Table 9.12

	A	B	C	D	E	F	G
1	*Groups*	*Count*	*Sum*	*Average*	*Variance*		
2	Column 1	6	9.784412	1.630735	0.045709		
3	Column 2	6	9.343754	1.557292	0.0187		
4	Column 3	6	8.780084	1.463347	0.019863		
5							
6	AnoVa						
7	*Source of variation*	*SS*	*df*	*MS*	*F*	*P-value*	*F crit*
8	Between groups	0.084477	2	0.042238	1.503633	0.253995	3.682317
9	Within groups	0.421362	15	0.028091			
10	Total	0.505839	17				

Exercise 31 🖑

Another blood sample was repeatedly assayed for α-foetoprotein as described in Exercise 30. Use a computer to analyse the raw data, decide if there is significant variation between blocks and, if appropriate, calculate an overall mean. Data given in Table 9.13 are in units of radioactivity bound (%).

Table 9.13

Block A	Block B	Block C
2.15	2.22	2.08
2.21	2.19	2.14
2.19	2.15	2.09
2.18	2.21	2.12
2.21	2.19	2.12
2.22	2.17	2.12

Relationships between variables

Exercise 32 ⧖

A series of dilutions of a standard solution of lactate dehydrogenase was assayed using a method similar to that described in Chapter 3. The assay results, given in Table 9.14, are in units of $\mu mol\ s^{-1}\ l^{-1}$. From the raw data plot a calibration curve (actually a straight line) with concentration in mass per unit volume on the x-axis and with the assay result on the y-axis. Determine the slope and the intercept on the y-axis of the line you fitted using a graphical method. (The extra columns in Table 9.14 are for the next exercise.)

Table 9.14

$\mu g\ ml^{-1}$	$\mu mol\ s^{-1}\ l^{-1}$	Predicted	Residual
0.00	0.07		
0.05	1.83		
0.10	2.56		
0.15	4.57		
0.20	6.58		
0.25	6.80		
0.30	9.33		
0.35	10.57		
0.40	11.05		
0.45	13.57		
0.50	15.58		

Exercise 33 ⧖

For each of the concentrations (in mass per unit volume) in the previous exercise, use your estimated slope and intercept to predict the reaction velocity. Then calculate the residual. Finally make a residual plot on graph paper, i.e. plot residual against concentration in mass per unit volume. Use judgement by eye to answer the following:

(a) Could you have made a better fit to the data?

(b) Are the data consistent with a linear relationship?

(c) Are the data consistent with their being homoschedastic?

Exercise 34 ✍

A number of equations follow. For each one decide whether it is possible to transform data to enable the use of linear regression. What must you use as the independent variable of the regression? What must you use as the dependent variable of the regression? How will you find the constants in the equation using the coefficients from the linear regression?

(a) $a = kb^3 + c$

Where a is the dependent variable, b is the independent variable, and k and c are constants.

(b) $a = \dfrac{1}{kb} + c$

Where a is the dependent variable, b is the independent variable, and k and c are constants.

(c) $a = k\dfrac{b}{1 - b}$

Where a is the dependent variable, b is the independent variable, and k is a constant.

Exercise 35

Continuing from Exercise 34, if you assume that:

1. the independent variable is determined with perfect precision;

2. variation in the dependent variable is normal; and

3. variation in the dependent variable is homoschedastic;

which transforms will allow valid non-weighted linear regression?

Exercise 36

Table 9.15 repeats the data from the calibration curve in Exercise 17.

Table 9.15		
[afp] ng ml^{-1}	Log$_{10}$([afp] ng ml^{-1})	Bound (%)
0.25		11.8
0.5		10
1		8
2.5		5
5		3.5
10		2

Calculate the base 10 logarithm of the α-foetoprotein concentrations. Now plot percentage-bound radioactivity (*y*-axis) against log of concentration (*x*-axis). Estimate the slope and intercept of this graph.

The result of assaying a serum sample was a value for bound radioactivity of 7%. Use your slope and intercept to calculate the concentration of α-foetoprotein in the serum sample in ng ml^{-1}.

Another result of assaying a serum sample was a value for bound radioactivity of 15%. Use your slope and intercept to calculate the concentration of α-foetoprotein in the serum sample in ng ml^{-1}. Is it reasonable to extrapolate from your calibration curve like this?

Exercise 37 ⌛

Repeat Exercise 36 but use a pocket calculator to apply linear regression. The following assumptions are made:

1. Since the standard solutions were produced by simple dilution we assume that the independent variable is determined with perfect precision. (If the concentrations are perfectly precise the logs of the concentrations will also be perfectly precise.)

2. There is a linear relationship between log of concentration of α-foetoprotein and percentage-bound radioactivity. This is a reasonable assumption for the range of values chosen but if the calibration curve were extended you would see a sigmoidal graph.

3. Random variation in percentage-bound radioactivity is normally distributed.

4. Precision of the determination of percentage-bound radioactivity is constant for high and low values.

Exercise 38 🖰

Carry out the same exercise a third time but this time use computer software. You should get identical results for the slope and intercept as the pocket calculator method produced. You will additionally get 95% confidence limits on the slope and intercept.

To visualize the confidence limits plot the following five lines on a graph:

(a) The regression line.

(b) A line defined by upper limit of intercept and upper limit of slope.

(c) A line defined by lower limit of intercept and upper limit of slope.

(d) A line defined by upper limit of intercept and lower limit of slope.

(e) A line defined by lower limit of intercept and lower limit of slope.

How does this match up with your hand-drawn graph from exercise 36? Does your 'by eye' line match well the regression line? Does the level of uncertainty suggested by the regression method confirm your subjective uncertainty over the graphical method?

Exercise 39

The initial rate of an enzyme-catalysed reaction is under study. The initial rate of consumption of the substrate, v_0 was determined for a range of different substrate concentrations, [S]. Replicate measurements of v_0 were made for each value of [S]. Non-linear, non-weighted regression was applied to the untransformed data with the assumption that classic Michaelis–Menton kinetics apply.

The regression gave the following estimates:

$K_m = 340$ μM

$V_{max} = 0.0441$ μmol s^{-1}

From the data and the regression coefficients, residuals were calculated and Figure 9.1 is the residual plot.

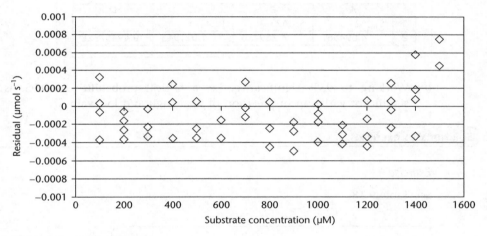

Figure 9.1   Residual plot.

(a) From the residual plot, use judgement by eye to test the following statements:
  (i) The data are consistent with an enzyme that conforms to classic Michaelis–Menton kinetics.
  (ii) The data are consistent with precision in v_0 being equal for high and low values.
  A statistical test like the non-linearity test described in Chapter 6 was applied to the data. The only difference from the non-linearity calculation was that predicted v_0 values were calculated using the Michealis–Menton equation rather than the equation of a straight line. Table 9.16 shows the figures produced.

(b) Working to 95% significance, i.e. a critical probability of 0.05, do the data significantly deviate from classic Michaelis–Menton kinetics?

An f_{max} test, as described in Chapter 6, was applied to the data. The results are given in Table 9.17.

Table 9.16

	A	B	C	D	E	F
1	AnoVa					
2	*Source of variation*	*SS*	*df*	*MS*	*F*	*P-value*
3	**Between groups**	10298.23	9	1144.248	900.4262	5.72585E–34
4	Regression	10255.64	1	10255.64		
5	Non-MM	42.58987	9	4.732208	3.723847	0.003092537
6	**Within groups**	38.12354	30	1.270785		
7						
8	Total	10336.35	39			

Table 9.17

	A	B	C	D
1	$S_{max}\text{^}2$	$S_{min}\text{^}2$	F_{max}	p
2	4.894561	4.789456	1.021945	0.49457328

(c) Working to 95% significance, i.e. a critical probability of 0.05, is there significant variation in precision of v_0 at high and low values?

Solutions to exercises

Exercise 1

See Table 9.18.

Table 9.18

Mathematical	Percentage	Integer ratio
0.33333	33.333%	1 in 3
0.25	25%	1 in 4
0.01	1%	1 in 100
0.0001	0.01%	1 in 10 000

Exercise 2

The probability that a baby born in Ruritania will have hypothyroidism is:

$$\frac{35}{210\,125} = 0.000167$$
$$= 0.0167\%$$

Exercise 3

The probability that a girl in the republic of Santa Cruz will be born with hypo-thyroidism is:

$$\frac{33}{152\,800} = 0.000216$$
$$= 0.0216\%$$

The probability that a boy in the republic of Santa Cruz will be born with hypo-thyroidism is:

$$\frac{17}{153\,205} = 0.000111$$
$$= 0.0111\%$$

Exercise 4

Points to consider in your discussion:

1. For the purposes of predicting the number of babies with hypothyroidism that will need treatment, the republic of Santa Cruz study was no more useful since all that is required is the overall percentage of babies that will be born with the disease.

2. Both studies produced valid probabilities although each result was different.

3. If the aim is to give pregnant women advice about the probability of their baby being found to have hypothyroidism on birth then the republic of Santa Cruz data is much more useful. A pregnant woman may have had an ultrascan and she may know the sex of her baby. Since the probability of a female baby having hypothyroidism is much higher than for a male, the extra information will allow doctors to change their risk assessment appropriately.

4. The studies done don't give any indication that race affects the probability that a child will be born with hypothyroidism. (In fact it does.) So it might be wise for a country to treat data from a study on a different ethnic population with caution.

Exercise 5

$$P(NOT\ H_0) = 1.0 - P(H_0)$$
$$= 1.0 - 0.025$$
$$= 0.975 = 97.5\%$$

Exercise 6

(a) The probability that a person will get a positive result with this faecal occult blood test is:

$$0.001\% + 0.999\% = 1\%$$

(b) The probability that a person will get a negative result with this faecal occult blood test is:

98.901% + 0.099% = 99%

Exercise 7

(a) The probability that a woman entering the clinic is taking steroid-based drugs is:

$$\frac{205}{675} = 0.3037$$

(b) The probability that a woman entering the clinic is not taking steroid-based drugs is:

1.0 − 0.3037 = 0.6963

(c) The probability that a woman entering the clinic is a regular smoker is:

$$\frac{278}{702} = 0.3960$$

(d) The probability that a woman entering the clinic is not a regular smoker is:

1.0 − 0.3960 = 0.6040

(e) Taking steroid drugs and is a smoker:

0.3037 × 0.3960 = 0.1203

(f) Taking steroid drugs and is not a smoker:

0.3037 × 0.6040 = 0.1834

(g) Not taking steroid drugs and is a smoker:

0.6963 × 0.3960 = 0.2757

(h) Not taking steroid drugs and is not a smoker:

0.6963 × 0.6040 = 0.4206

(i) The study has a serious flaw because the survey on use of steroid drugs and the survey on smoking were conducted separately. This forces us to use mathematics that uses independent combinations of probabilities. The problem with this is that women who use steroid-based drugs may already be paying good attention to health advice and they may be less likely to be smokers. This won't be detected by the experimental design used here.

Exercise 8

(a) $P(A) = \dfrac{354}{98\,456} = 0.003596$

(b) $P(NOT\ A) = 1.0 - P(A) = 0.996404$

(c) $P(B|A) = \dfrac{55}{256} = 0.214844$

(d) $P(\text{NOT } B | A) = \dfrac{201}{256} = 0.785156$

(e) $P(B | \text{NOT } A) = \dfrac{100}{1032} = 0.096899$

(f) $P(\text{NOT } B | \text{NOT } A) = \dfrac{932}{1032} = 0.903101$

(g) $P(A \text{ AND } B) = P(A) \times P(B | A) = 0.000772$

(h) $P(A \text{ AND NOT } B) = P(A) \times P(\text{NOT } B | A) = 0.002823$

(i) $P(\text{NOT } A \text{ AND } B) = P(\text{NOT } A) \times P(B | \text{NOT } A) = 0.096551$

(j) $P(\text{NOT } A \text{ AND NOT } B) = P(\text{NOT } A) \times P(\text{NOT } B | \text{NOT } A) = 0.899854$

Why have we ended up with ten different probability figures? It's because we have taken samples of data to try and estimate probabilities for three different groups of people:

1. All people in the population who may or may not use the blood-pressure drug and who may or may not use the nasal spray. There are four different permutations of using/not using the nasal spray/blood-pressure drug (Table 9.19).

Table 9.19			
	B	**NOT B**	**Total**
A	P(A AND B)	P(A AND NOT B)	P(A)
NOT A	P(NOT A AND B)	P(NOT A AND NOT B)	P(NOT A)
Total	P(B)	P(NOT B)	1.0

2. All people who definitely use the blood-pressure drug but may or may not use the nasal spray (Table 9.20).

Table 9.20			
	B	**NOT B**	**Total**
A	P(B|A)	P(NOT B|A)	1

3. All people who definitely do *not* use the blood-pressure drug but may or may not use the nasal spray (Table 9.21).

Table 9.21			
	B	**NOT B**	**Total**
NOT A	P(B|NOT A)	P(NOT B|NOT A)	1

Exercise 9

Your safest course in explaining the risk may simply be to describe the study that was done. For example: 'This is a very safe operation but there is some risk and I'm able to give you some reassurance because records have been kept locally going back for some years. The records show that the operation has been done on more than 7000 patients and we have had only 23 fatalities. I can't be completely certain that your son will not be harmed by the operation but the risk that he might die as a result is very low.'

Exercise 10

There is no easy answer to this one. You have to tread a fine line between getting across the point that there is always some risk involved in operations and sounding like you are just trying to hide behind figures and statistics.

Exercise 11

It is possible to isolate and purify the peroxidase enzyme from horseradish and so you could design an experiment where the volunteers are grouped into a control group with no horseradish or its peroxidase in the diet and a group taking purified horseradish peroxidase. If horseradish sauce has an effect but the peroxidase does not then you might look at other components of the sauce.

Exercise 12

The calculation of expected frequencies is shown in Table 9.22.

Table 9.22

Expected	Using drug	Not taking drug	Total
Using spray	$256 \times \frac{155}{1288} = 30.81$	$1032 \times \frac{155}{1288} = 124.19$	155
Not using spray	$256 \times \frac{1133}{1288} = 225.19$	$1032 \times \frac{1133}{1288} = 907.81$	1133
Total	256	1032	1288

Exercise 13

$\chi^2 = 25.85$

Exercise 14

$p = 3.69 \times 10^{-7}$

$= 0.0000369\%$

Exercise 15

You may have laid out your spreadsheet differently, but compare your calculations to Table 9.23.

Table 9.23

	A	B	C
1		Volumes	
2	Serum	0.05	ml
3	NADH	2.5	ml
4	Pyruvate	0.5	ml
5	Total	=SUM(B2:B4)	
6			
7	MolExtCo		
8	6.23	ml per micromol per cm	
9			
10	Initial A	After A	Delta A
11	1.02	0.98	=A11-B11
12			
13	micromols consumed		
14	=B5*C11/A8		
15			
16	Time		
17	180		
18			
19	micromols per sec		
20	=A14/A17		
21			
22	micromol per second per litre serum		
23	=A20/(B2/1000)		

Exercise 16

See Table 9.24.

Table 9.24

Calculation	Units
$\dfrac{0.05 \times 3.0}{5.0 \times 1.0}$	μmol
$\dfrac{5.0 \times 3.0}{0.05 \times 1.0}$	$ml^2\,\mu mol^{-1}\,cm^{-2}$
$\dfrac{0.05 \times 1.0}{5.0 \times 3.0}$	$ml^{-2}\,\mu mol\,cm^2$
$\dfrac{0.05 \times 3.0 \times 1.0}{5.0}$	$\mu mol\,cm^2$

Exercise 17

Compare your hand-drawn graph with the graph in Figure 9.2.

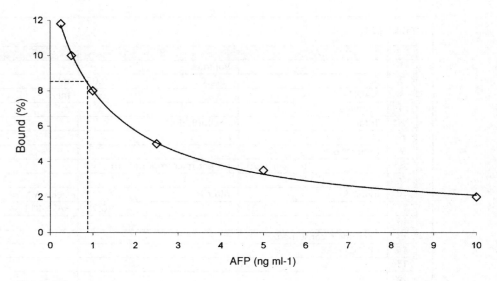

Figure 9.2

Graphical interpolation gives an α-foetoprotein concentration of about 0.8 ng ml^{-1} for experimental radioactivity 8.5 bound (%).

Exercise 18

In the second set of data the following statistical values will be lower than in the first set of data:

(a) The mean, because it is nearer the correct value.

(b) The lower confidence limit of the mean, because precision is the same and it will stay the same distance from the mean and drop with it.

(c) The upper confidence limit of the mean, for the same reason.

The following statistical values will stay roughly the same although they may vary due to random variation. You can't predict that they will drop:

(d) The standard error of the mean.

(e) The standard deviation.

Exercise 19

(a) 3.127×10^3

(b) 7.91×10^{-3}

(c) $6.023{\times}10^{23} \times 1.0{\times}10^{-27}$ $= 6.023 \times 1.0 \times 10^{-27} \times 10^{23}$

$$= 6.023 \times 10^{(23-27)}$$

$$= 6.023{\times}10^{-4}$$

(d) $(2.0{\times}10^{-5} \times 4.2{\times}10^2) + 4.5 \times 10^{-4}$ $= (2.0 \times 4.2 \times 10^{(2-5)}) + 4.5 \times 10^{-4}$

$$= 8.4 \times 10^{-3} + 0.45 \times 10^{-3}$$

$$= 8.85{\times}10^{-3}$$

(e) $3.054{\times}10^{-6} + 5.78{\times}10^{-7}$ $= 3.054{\times}10^{-6} + 0.578{\times}10^{-6}$

$$= 3.632 \times 10^{-6}$$

(f) $2.23{\times}10^{16} - 8.8{\times}10^{14}$ $= 2.23{\times}10^{16} - 0.088{\times}10^{16}$

$$= 2.142{\times}10^{16}$$

(g) $\dfrac{1.2{\times}10^{-16}}{4.0{\times}10^{-2}} = \dfrac{1.2}{4.0} \times 10^{-16} \times 10^2$

$$= 0.3 \times 10^{(2-16)}$$

$$= 0.3 \times 10^{-14}$$

$$= 3{\times}10^{-13}$$

(h) $\dfrac{3.0{\times}10^{-5} + 5.5{\times}10^3}{5.0{\times}10^{17}} = (0.00000003 \times 10^3 + 5.5 \times 10^3) \times \dfrac{1}{5} \times 10^{-17}$

$$= 5.50000003 \times 0.2 \times 10^{3-17}$$

$$= 1.100000006{\times}10^{-14}$$

Exercise 20

(a) 5^6 (b) 3^6 (c) x (d) $3x^{-6}$ (e) y^2x^3

(f) $x^{\left(\frac{1}{3}\times 6\right)} = x^2$ (g) $x^2 \times x^{0.5} \times x^{-3.5} = x^{-1} = \frac{1}{x}$

Exercise 21

(a) Since $3^3 = 27$, $\log_3 27 = 3$ so $x = 3$

(b) $x = 16^{1/2} = \sqrt{16} = 4$

(c) $8 = x^{1/2} = \sqrt{x}$ so $x = 8^2 = 64$

(d) $x = b^y$

(e) $x^3 = b^y$ so $x = \sqrt[3]{b^y} = b^{y/3}$

(f) $x^y = 3$ so $x = \sqrt[y]{3} = 3^{1/y}$

(g) $\sqrt{x^3} = b^y$ so $x = b^{2y/3}$

(h) $\log_b 6 = \log_b(2 \times 3)$

$$= \log_b 2 + \log_b 3$$

$$= 0.69 + 1.1 = 1.79$$

(i) $\log_b\left[3 \times \sqrt[2]{3}\right] = \log_b 3 + \dfrac{(\log_b 3)}{2}$

$= 1.1 \times 1.5 = 1.65$

(j) $\log_b\left[12^{1/4}\right] = \dfrac{(\log_b 12)}{4}$

$= \dfrac{\log_b(2 \times 3 \times 3)}{4} = \dfrac{(0.69 + 1.1 + 1.1)}{4} = 0.7225$

(k) $\log_c 9 = \log_c(3 \times 3)$

$= 2 \times \log_c 3$

$= 2 \times 0.95 = 1.9$

(l) $\log_c 16 = \log_c(4 \times 4)$

$= 2 \times \log_c 4$

$= 2 \times 1.20 = 2.4$

(m) $\log_c 12 = \log_c(3 \times 4)$

$= \log_c 3 + \log_c 4$

$= 0.95 + 1.20 = 2.15$

(n) $\log_c(3^3 \times 4^6) = 3 \times \log_c 3 + 6 \times \log_c 4$

$= 3 \times 0.95 + 6 \times 1.2 = 7.2$

Exercise 22

According to the normal distribution, 95% of measurements lie between 1.960 standard deviations of the mean so:

$X_{\text{upper}} = 10 + (0.2 \times 1.960)$

$= 10.392$ bound (%)

$X_{\text{lower}} = 10 - (0.2 \times 1.960)$

$= 9.608$ bound (%)

Exercise 23

Sample mean = 4.98 bound (%)

Sample standard deviation = 0.217 bound (%)

Exercise 24

(a)	Mean	3.48	bound (%)
	s	0.0779	bound (%)
(b)	SE	0.0235	bound (%)
(c)	X_{upper}	3.53	bound (%)
(d)	X_{lower}	3.43	bound (%)

(e) Estimated concentration = 3.48 bound (%). The true value lies between 3.53 bound (%) and 3.43 bound (%) with 95% confidence.

Exercise 25

Figure 9.3 is a calibration curve with the two data points and their confidence limits interpolated.

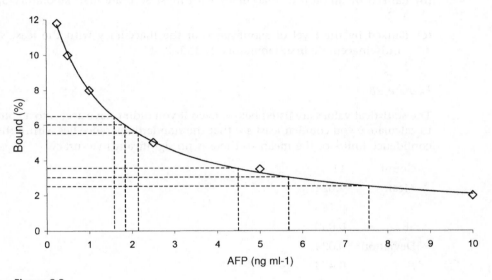

Figure 9.3

The first measurement = 1.8 ng ml^{-1} 95% confidence limits, between 1.6 and 2.1 ng ml^{-1}.

The second measurement = 5.7 ng ml^{-1} 95% confidence limits, between 4.5 and 7.6 ng ml^{-1}.

You should notice that precision in mass per unit volume is poorer at higher concentrations and that the confidence limits are not symmetrical either side of the measurements. This suggests that the assay is not only heteroschedastic but the imprecision is not normally distributed either.

Exercise 26

The consequences of important improvements in the precision of α-foetoprotein assays was discussed at the start of Chapter 4. If improper procedures were used to implement the normally precise protocols for α-foetoprotein then all the advantages of precision will be lost. Worse than that, if clinicians believe that the results they see are precise they may make medical decision that are wrong or may give advice to pregnant women that is false. For example, if an assay run for a blood sample from a pregnant woman contained a random error that was large and positive the doctor may incorrectly advise the woman that there is a high probability of her child having spina bifida and she may then choose to have an abortion.

Exercise 27

(a) Ranked by precision with best precision first: laboratory 5, 3, 4, 2, 1.

(b) Ranked by amount of inaccuracy with most accurate first: laboratory 3, 1, 5, 4, 2.

(c) Ranked by the level of significance of the inaccuracy with the least significantly inaccurate first: laboratory 1, 3, 5, 2, 4.

Exercise 28

The statistical values are listed below. Even if you didn't have access to a computer to calculate p you could at least see that the standard reference lies within the 95% confidence limits of the mean so there is no significant inaccuracy.

Count	11
Mean	4.024
s	0.166
SE	0.050
Deviation	0.024
t	0.473
$t_{0.05}$	2.228
X_{upper}	4.135
X_{lower}	3.912

Exercise 29

(a) Using the Excel formula TDIST(s, 10, 2) you will find $p = 0.646 = 64.6\%$ which confirms that the deviation of the mean from the reference value is not at all significant and easily explicable by imprecision.

(b) The measured concentration differs from the reference standard by 0.024 μmol s^{-1} l^{-1}. The probability that this is because of random variation in the individual measurements alone is 65%. The probability that it is owing partly to inaccuracy is 35%.

Exercise 30

Although the printout includes a number of figures only one is relevant to the question asked and that is the p value. Since $p = 0.25$ and this is much greater than the critical probability of 0.05 we conclude that there is no significant variation between block means so it is appropriate to pool the data and calculate the mean:

Mean = 1.55 ng ml^{-1}

(Calculated by summing the three block sums and dividing by 18.)

Exercise 31

Table 9.25 gives the results you will obtain from Excel '97. You can see that a highly significant difference between block means has been detected and it is not appropriate to pool the data for a mean.

Table 9.25

	A	B	C	D	E	F	G
1	SUMMARY						
2	*Groups*	*Count*	*Sum*	*Average*	*Variance*		
3	Column 1	6	13.16	2.193333	0.000667		
4	Column 2	6	13.13	2.188333	0.000657		
5	Column 3	6	12.67	2.111667	0.000497		
6							
7	AnoVa						
8	*Source of variation*	*SS*	*df*	*MS*	*F*	*P-value*	*F crit*
9	Between groups	0.025144	2	0.012572	20.72344	4.82E–05	3.682317
10	Within groups	0.0091	15	0.000607			
11							
12	Total	0.034244	17				

Exercise 32

Your hand-drawn graph should look something like Figure 9.4. Your slope should be close to 30.0 μmol s^{-1} l^{-1} μg^{-1} ml. Your intercept on the y-axis should be close to zero.

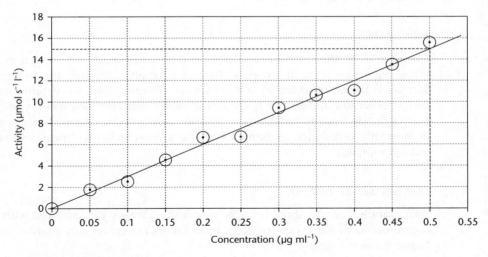

Figure 9.4 Calibration curve.

Exercise 33

If you effectively fitted a line to the original data your residual plot should look similar to Figure 9.5.

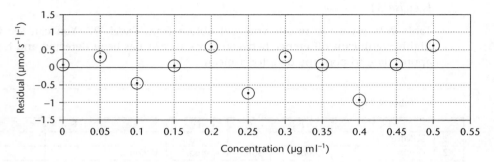

Figure 9.5 Calibration curve – residual plot.

If the points on your residual plot are not roughly spread across the four quadrants of the graph then you may not have fit a good line to the data. From Figure 9.5 you can see the data are consistent with a linear relationship and with the assay being heteroschedastic.

Exercise 34

(a) Plot a against b^3. The slope of the graph will be k and the intercept on the y-axis will be c.

(b) Plot a against $1/b$. The slope of the graph will be $1/k$ and the intercept on the y-axis will be c.

(c) Plot $1/a$ against $1/b$. The slope of the graph will be $1/k$ and the intercept on the y-axis will also be $1/k$.

Exercise 35

For (a) and (b) the dependent variable is not transformed and so non-weighted linear regression is an appropriate method to apply. Although the independent variable is transformed for (a) and (b) we are assuming that it is determined with perfect precision and so the transformed values will also have perfect precision. For (c) the dependent variable is transformed and the transformed values cannot be normally distributed or homoschedastic so non-weighted linear regression is not appropriate.

Exercises 36, 37 and 38

Your graphical results should closely match the answers you achieved with linear regression. Table 9.26 is the printout from Excel's regression data analysis tool and Figure 9.6 is the graph.

Table 9.26

	A	B	C	D	E	F	G
1	SUMMARY OUTPUT						
2							
3	*Regression statistics*						
4	Multiple R	0.99646236					
5	R Square	0.995298012					
6	Adjusted R Square	0.994122515					
7	Standard Error	0.294553012					
8	Observations	6					
9							
10	ANOVA						
11		*df*	*SS*	*MS*	*F*	*Significance F*	
12	Regression	1	73.46128743	73.46128743	846.7039739	8.30379E-06	
13	Residual	4	0.347045908	0.08676147			
14	Total	5	73.80833333				
15							
16		*Coefficients*	*Standard Error*	*t Stat*	*P-value*	*Lower 95%*	*Upper 95%*
17	Intercept	7.96268079	0.127688921	62.38809131	3.95367E-07	7.611746064	8.320790093
18	X Variable 1	−6.28035073	0.215833125	−29.09817819	8.30379E-06	−6.879600795	−5.681100665

Interpolation: 7 bound (%) corresponds to 1.43 ng ml^{-1}

Extrapolation: 15 bound (%) corresponds to 0.0759 ng ml^{-1}

It is very dangerous to extrapolate because the transformed data may not be linear beyond the range of concentrations used in the experiment. In fact, the graph would be distinctly sigmoidal if lower and higher concentrations had been used. The graph, with various plausible regression lines is shown in Figure 9.6. This is not the statistically correct way to put confidence limits on the regression line but it's a quick and simple way to show just how much uncertainty there is over the slope and intercept produced by linear regression.

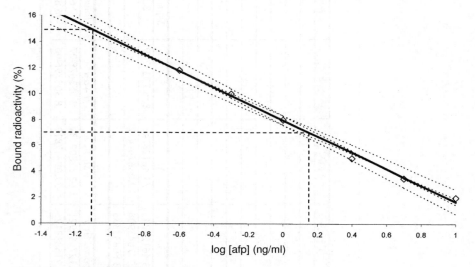

Figure 9.6   AFP calibration curve.

Exercise 39

(a) This exercise was just about picking out the appropriate statistical values from the computer output. Judging the residual plot by eye doesn't lead to clear-cut conclusions. It looks a little like residuals have a tendency to be positive on the right-hand edge of the graph but the residuals seem to be more or less spread the same amount for high and low substrate concentrations.

(b) The probability that the data deviate from the Michaelis–Menton kinetics owing only to imprecision in velocity measurements is 0.00309, i.e. there is a highly significant deviation.

(c) The probability that the variation in velocity measurements significantly varies for different substrate concentrations owing only to random imprecision is 0.495, i.e. there is no significant heteroschedasticity.

Index

LIVERPOOL
JOHN MOORES UNIVERSITY
AVRIL ROBARTS LRC
TITHEBARN STREET
LIVERPOOL L2 2ER
TEL. 0151 231 4022

Data analysis
for biomolecular
sciences

LIVERPOOL
JOHN MOORES UNIVERSITY
AVRIL ROBARTS LRC
TITHEBARN STREET
LIVERPOOL L2 2ER
TEL. 0151 231 4022

LIVERPOOL JMU LIBRARY

3 1111 00812 1640